CALCULUS MADE EASY

BOOKS BY MARTIN GARDNER

Fads and Fallacies in the Name of
Science

Mathematics, Magic, and Mystery

Great Essays in Science (ed.)

Logic Machines and Diagrams

The Scientific American Book of
Mathematical Puzzles and Diversions

The Annotated Alice

The Second Scientific American Book of
Mathematical Puzzles and Diversions

Relativity for the Million

The Annotated Snark

The Ambidextrous Universe

The Annotated Ancient Mariner

New Mathematical Diversions from
Scientific American

The Annotated Casey at the Bat

Perplexing Puzzles and Tantalizing
Teasers

The Unexpected Hanging and Other
Mathematical Diversions

Never Make Fun of a Turtle, My Son
(verse)

The Sixth Book of Mathematical Games
from Scientific American

Codes, Ciphers, and Secret Writing

Space Puzzles

The Snark Puzzle Book

The Flight of Peter Fromm (novel)

Mathematical Magic Show

More Perplexing Puzzles and
Tantalizing Teasers

The Encyclopedia of Impromptu Magic

Aha! Insight

Mathematical Carnival

Science: Good, Bad, and Bogus

Science Fiction Puzzle Tales

Aha! Gotcha

Wheels, Life, and Other Mathematical
Amusements

Order and Surprise

The Whys of a Philosophical Scrivener

Puzzles from Other Worlds

The Magic Numbers of Dr. Matrix

Knotted Doughnuts and Other
Mathematical Entertainments

The Wreck of the Titanic Foretold

Riddles of the Sphinx

The Annotated Innocence of Father
Brown

The No-Sided Professor (short stories)

Time Travel and Other Mathematical
Bewilderments

The New Age: Notes of a Fringe
Watcher

Gardner's Whys and Wherefores

Penrose Tiles to Trapdoor Ciphers

How Not to Test a Psychic

The New Ambidextrous Universe

More Annotated Alice

The Annotated Night Before Christmas

Best Remembered Poems (ed.)

Fractal Music, Hypercards, and More

The Healing Revelations of Mary Baker
Eddy

Martin Gardner Presents

My Best Mathematical and Logic
Puzzles

Classic Brainteasers

Famous Poems of Bygone Days (ed.)

Urantia: The Great Cult Mystery

The Universe Inside a Handkerchief

The Night Is Large

Last Recreations

Visitors from Oz

CALCULUS MADE EASY

BEING A VERY-SIMPLEST INTRODUCTION TO THOSE
BEAUTIFUL METHODS OF RECKONING WHICH
ARE GENERALLY CALLED BY THE
TERRIFYING NAMES
OF THE

DIFFERENTIAL CALCULUS

AND THE

INTEGRAL CALCULUS

Silvanus P. Thompson, F.R.S.

AND

Martin Gardner

Newly Revised, Updated, Expanded, and
Annotated for its 1998 edition.

ST. MARTIN'S PRESS

New York

The original edition of *Calculus Made Easy* was written by Silvanus P.
Thompson and published in 1910, with subsequent editions in 1914 and
1946.

Production Editor: David Stanford Burr
Design: Susan Hood

Library of Congress Cataloging–in–Publication Data

Thompson, Silvanus Phillips, 1851–1916.
    Calculus made easy : being a very-simplest introduction to those
beautiful methods of reckoning which are generally called by the
terrifying names of the differential calculus and the integral
calculus. — Newly rev., updated, expanded, and annotated for its
1998 ed. / Silvanus P. Thompson and Martin Gardner.
        p.   cm.
    ISBN 0-312-18548-0
    1. Calculus.   I. Gardner, Martin.   II. Title.
QA303.T45   1998
515—dc21   98-10433
                                                        CIP

10   9   8   7

CONTENTS

PREFACE TO THE 1998 EDITION

Introductory courses in calculus are now routinely taught to high school students and college freshmen. For students who hope to become mathematicians or to enter professions that require a knowledge of calculus, such courses are the highest hurdle they have to jump. Studies show that almost half of college freshmen who take a course in calculus fail to pass. Those who fail almost always abandon plans to major in mathematics, physics, or engineering—three fields where advanced calculus is essential. They may even decide against entering such professions as architecture, the behavioral sciences, or the social sciences (especially economics) where calculus can be useful. They exit what they fear will be too difficult a road to consider careers where entrance roads are easier.

One reason for such a high dropout rate is that introductory calculus is so poorly taught. Classes tend to be so boring that students sometimes fall asleep. Calculus textbooks get fatter and fatter every year, with more multicolor overlays, computer graphics, and photographs of eminent mathematicians (starting with Newton and Leibniz), yet they never seem easier to comprehend. You look through them in vain for simple, clear exposition and for problems that will hook a student's interest. Their exercises have, as one mathematician recently put it, "the dignity of solving crossword puzzles." Modern calculus textbooks often contain more than a thousand pages—heavy enough to make excellent doorstops—and more than a thousand frightening exercises! Their prices are rapidly approaching $100.

"Why do calculus books weigh so much?" Lynn Arthur Steen

asked in a paper on "Twenty Questions for Calculus Reformers" that is reprinted in *Toward a Lean and Lively Calculus* (Mathematical Association of America, 1986), edited by Ronald Douglas. Because, he answers, "the economics of publishing compels authors . . . to add every topic that anyone might want so that no one can reject the book just because some particular item is omitted. The result is an encyclopaedic compendium of techniques, examples, exercises and problems that more resemble an overgrown workbook than an intellectually stimulating introduction to a magnificent subject."

"The teaching of calculus is a national disgrace," Steen, a mathematician at St. Olaf College, later declared. "Too often calculus is taught by inexperienced instructors to ill-prepared students in an environment with insufficient feedback."

Leonard Gillman, writing on "The College Teaching Scandal" (*Focus,* Vol. 8, 1988, page 5), said: "The calculus scene has been execrable for many years, and given the inertia of our profession is quite capable of continuing that way for many more."

Calculus has been called the topic mathematicians most love to hate. One hopes this is true only of teachers who do not appreciate its enormous power and beauty. Howard Eves is a retired mathematician who actually enjoyed teaching calculus. In his book *Great Moments in Mathematics* I found this paragraph:

> Surely no subject in early college mathematics is more exciting or more fun to teach than the calculus. It is like being the ringmaster of a great three-ring circus. It has been said that one can recognize the students on a college campus who have studied the calculus—they are the students with no eyebrows. In utter astonishment at the incredible applicability of the subject, the eyebrows of the calculus students have receded higher and higher and finally vanished over the backs of their heads.

Recent years have seen a great hue and cry in mathematical circles over ways to improve calculus teaching. Endless conferences have been held, many funded by the federal government. Dozens of experimental programs are underway here and there. Some leaders of reform argue that while traditional textbooks get weightier, the need for advanced calculus is actually diminishing. In his popular *Introduction to the History of Mathematics,* Eves sadly writes: "Today the larger part of mathematics has no, or very little connection with calculus or its extensions."

Why is this? One reason is obvious. Computers! Today's digital computers have become incredibly fast and powerful. Continuous functions which once could be handled only by slow analog machines can now be turned into discrete functions which digital computers handle efficiently with step-by-step algorithms. Hand-held calculators called "graphers" will instantly graph a function much too complex to draw with a pencil on graph paper. The trend now is away from continuous math to what used to be called finite math, but now is more often called discrete math.

Calculus is steadily being downgraded to make room for combinatorics, graph theory, topology, knot theory, group theory, matrix theory, number theory, logic, statistics, computer science, and a raft of other fields in which continuity plays a relatively minor role.

Discrete mathematics is all over the scene, not only in mathematics but also in science and technology. Quantum theory is riddled with it. Even space and time may turn out to be quantized. Evolution operates by discrete mutation leaps. Television is on the verge of replacing continuous analog transmission by discrete digital transmission which greatly improves picture quality. The most accurate way to preserve a painting or a symphony is by converting it to discrete numbers which last forever without deteriorating.

When I was in high school I had to master a pencil-and-paper way to calculate square roots. Happily, I was not forced to learn how to find cube and higher roots! Today it would be difficult to locate mathematicians who can even recall how to calculate a square root. And why should they? They can find the *n*th root of any number by pushing keys in less time than it would take to consult a book with tables of roots. Logarithms, once used for multiplying huge numbers, have become as obsolete as slide rules.

Something similar is happening with calculus. Students see no reason today why they should master tedious ways of differentiating and integrating by hand when a computer will do the job as rapidly as it will calculate roots or multiply and divide large numbers. Mathematica, a widely used software system developed by Stephen Wolfram, for example, will instantly differentiate and integrate, and draw relevant graphs, for any calculus problem likely to arise in mathematics or science. Calculators with keys for finding derivatives and integrals now cost less than most calculus textbooks. It has been estimated that more than ninety percent of the exercises in the big textbooks can be solved by using such calculators.

Leaders of calculus reform are not suggesting that calculus no longer be taught, what they recommend is a shift of emphasis from problem solving, which computers can do so much faster and more accurately, to an emphasis on understanding what computers are doing when they answer calculus questions. A knowledge of calculus is even essential just to know what to ask a computer to do. Above all, calculus courses should instill in students an awareness of the great richness and elegance of calculus.

Although suggestions are plentiful for ways to improve calculus teaching, a general consensus is yet to emerge. Several mathematicians have proposed introducing integral calculus before differential calculus. A notable example is the classic two-volume *Differential and Integral Calculus* (1936–37) by Richard Courant. However, differentiating is so much easier to master than integrating that this switch has not caught on.

Several calculus reformers, notably Thomas W. Tucker (See his "Rethinking Rigor in Calculus," in *American Mathematical Monthly,* (Vol. 104, March 1997, pp. 231–240) have recommended that calculus texts replace the important mean value theorem (MVT) with an increasing function theorem (IFT). (On the mean value theorem see my Postscript to Thompson's Chapter 10.) The IFT states that if the derivative on a function's interval is equal to or greater than zero, then the function is increasing on that interval. For example, if a car's speedometer always shows a number equal to or greater than zero, during a specified interval of time, then during that interval the car is either standing still or moving forward. Stated geometrically, it says that if the curve of a continuous function, during a given interval, has a tangent that is either horizontal or sloping upward, the function on that interval is either unchanging or increasing. This change also has not caught on.

Many reformers want to replace the artificial problems in traditional textbooks with problems about applications of calculus in probability theory, statistics, and in the biological and social sciences. Unfortunately, for beginning students not yet working in these fields, such "practical" problems can seem as dull and tedious as the artificial ones.

More radical reformers believe that calculus should no longer be taught in high school, and not even to college freshmen unless they have decided on a career for which a knowledge of calculus is required. And there are opponents of reform who find nothing wrong in the way calculus has been traditionally taught, assuming, of course, it is taught by competent teachers.

In its February 28, 1992, issue, *Science* investigated Project Calc, a computer oriented calculus course offered at Duke University. Only 57 percent of its students continued on with a second course in calculus as compared with 68 percent who continued on after taking a more traditional course. A few students liked the experimental course. Most did not. One student called it "the worst class I ever took." Another described it as "a big exercise in confusion." Still another is quoted in *Science* as saying: "I am very jealous of my friends in normal calculus. I would do anything to have taken a regular calculus with pencil and paper."

Efforts are now underway to combine continuous and discrete mathematics in a single textbook. An outstanding example is *Concrete Mathematics* (1984, revised 1989), an entertaining textbook by Ronald Graham, Donald Knuth, and Oren Patashnik. The authors coined the term "concrete" by taking "con" from the start of "continuous," and "crete" from the end of "discrete." However, even this exciting textbook presupposes a knowledge of calculus.

The American philosopher and psychologist William James, in an 1893 letter to Theódore Flournoy, a Geneva psychologist, asked "Can you name me any simple book on the differential calculus which gives an insight into the philosophy of the subject?"

In spite of the current turmoil over fresh ways to teach calculus, I know of no book that so well meets James's request as the book you are now holding. Many similar efforts have been made, with such titles as *Calculus for the Practical Man, The ABC of Calculus, What Is Calculus About?, Calculus the Easy Way,* and *Simplified Calculus.* They tend to be either too elementary, or too advanced. Thompson strikes a happy medium. It is true that his book is old-fashioned, intuitive, and traditionally oriented. Yet no author has written about calculus with greater clarity and humor. Thompson not only explains the "philosophy of the subject," he also teaches his readers how to differentiate and integrate simple functions.

Silvanus Phillips Thompson was born in 1851, the son of a school teacher in York, England. From 1885 until his death in 1916 he was professor of physics at the City and Guilds Technical College, in Finsbury. A distinguished electrical engineer, he was elected to the Royal Society in 1891, and served as president of several scientific societies.

Thompson wrote numerous technical books and manuals on electricity, magnetism, dynamos, and optics, many of which went through several editions. He also authored popular biographies of scientists Michael Faraday, Philipp Reis, and Lord Kelvin. A devout Quaker and an active Knight Templar, he wrote two books about his faith: *The Quest for Truth* (1915), and *A Not Impossible Religion* (published posthumously 1918). He was much in demand as a lecturer, and said to be a skillful painter of landscapes. He also wrote poetry. In 1920 two of his four daughters, Jane Smeal Thompson and Helen G. Thompson, published a book about their father titled *Silvanus Phillips Thompson: His Life and Letters*.

Calculus Made Easy was first issued by Macmillan in England in 1910 under the pseudonym of F.R.S.—initials that stand for Fellow of the Royal Society. The identity of the author was not revealed until after his death. The book was reprinted three times before the end of 1910. Thompson revised the book considerably in 1914, correcting errors and adding new material. The book was further revised and enlarged posthumously in 1919, and again in 1945 by F. G. W. Brown. Some of these later additions, such as the chapter on partial fractions, are more technical than chapters in Thompson's original work. Curiously, Thompson's first edition, with its great simplicity and clarity, is in a way closer to the kind of introductory book recommended today by reformers who wish to emphasize the basic ideas of calculus, and to downplay tedious techniques for solving problems that today can be solved quickly by computers. Readers who wish merely to grasp the essentials of calculus can skip the more technical chapters, and need not struggle with solving all the exercises. The book has never been out of print. St. Martin's Press published its paperback edition in 1970.

Almost all reviews of the book's first edition were favorable. A reviewer for *The Athenaeum* wrote:

> It is not often that it falls to the lot of a reviewer of mathematical literature to read such a gay and boisterous book as this "very simplest introduction to those beautiful methods of reckoning which are generally called by the terrifying names of the differential calculus and the integral calculus." As a matter of fact, professional mathematicians will give a warm welcome to a book which is so orthodox in its teaching and so vigorous in its exposition.

Professor E. G. Coker, a colleague, said in a letter to Thompson:

> I am very pleased to hear that your little book on the calculus is likely to be available for general use. As you know, I have been teaching the elements of this subject to the junior classes here for some years, and I do not know of any other book so well adapted to give fundamental ideas. One of the great merits of the book is that it dispels the mysteries with which professional mathematicians envelope the subject. I feel sure that your little book, with its common sense way of dealing with elementary ideas of the calculus, will be a great success.

Many of today's eminent mathematicians and scientists first learned calculus from Thompson's book. Morris Kline, himself the author of a massive work on calculus, always recommended it as the best book to give a high school student who wants to learn calculus. The late economist and statistician Julian Simon, sent me his paper, not yet published, titled "Why Johnnies (and Maybe You) Hate Math and Statistics." It contains high praise for Thompson's book:

> *Calculus Made Easy* has been damned by every professional mathematician I have asked about it. So far as I know, it is

not used in any calculus courses anywhere. Nevertheless, al-most a century after its first publication, it still sells briskly in paperback even in college bookstores. It teaches a system of approximation that makes quite clear the central idea of calculus—the idea that is extraordinarily difficult to com-prehend using the mathematician's elegant method of limits.

Later on Simon asks:

Question: why don't high school and college kids get to learn calculus the Thompson way? Answer: Thompson's sys-tem has an unremediable fatal flaw: It is ugly in the eyes of the world-class mathematicians who set the standards for the way mathematics is taught all down the line; the run-of-the-mill college and high school teachers, and ultimately their students, are subject to this hegemony of the aesthetic tastes of the great. Thompson simply avoids the deductive devices that enthrall mathematicians with their beauty and elegance.

I mentioned earlier a book titled *Toward a Lean and Lively Cal-culus.* It contains papers by mathematicians who participated in a 1986 conference at Tulane University on how to improve calculus teaching. Most of the contributors urged cutting down on tech-niques for problem solving, stressing the understanding of ideas, integrating calculus with the use of calculators, and scaling down textbooks to leaner and livelier forms. Now, the leanest and live-liest introduction to calculus ever written is Thompson's *Calculus Made Easy,* yet Peter Renz was the only mathematician at the conference who had the courage to praise the book and list it as a reference.

The two most important concepts in calculus are functions and limits. Because Thompson more or less assumes that his readers understand both notions, I have tried in two preliminary chapters to make clearer what they both mean. And I have added a brief chapter about derivatives. Here and there throughout *Calculus Made Easy* I have inserted footnotes where I think something of interest can be said about the text. These notes are initialed M.G. to distinguish them from Thompson's notes.

Where Thompson speaks of British currency I have changed the values to dollars and cents. Terminology has been updated. Thompson uses the obsolete term "differential coefficient." I have changed it to "derivative." The term "indefinite integral" is still used, but it is rapidly giving way to "antiderivative," so I have made this substitution.

Thompson followed the British practice of raising a decimal point to where it is easily confused with the dot that stands for multiplication. I have lowered every such point to conform to American custom. Where Thompson used a now discarded sign for factorials, I have changed it to the familiar exclamation mark. Where Thompson used the Greek letter for epsilon, I have changed it to the english e. Where Thompson used the symbol $\log_e$, I have replaced it with ln. Finally, in a lengthy appendix, I have thrown together a variety of calculus-related problems that have a recreational flavor.

I hope my revisions and additions for this newly revised edition of *Calculus Made Easy* will render it even easier to understand—not just for high school and college students, but also for older laymen who, like William James, long to know what calculus is all about. Most mathematics deals with static objects such as circles and triangles and numbers. But the great universe "out there," not made by us, is in a constant state of what Newton called flux. At every microsecond it changes magically into something different.

Calculus is the mathematics of change. If you are not a mathematician or scientist, or don't intend to become one, there is no need for you to master the techniques for solving calculus problems by hand. But if you avoid acquiring some insight into the essentials of calculus, into what James called its philosophy, you will miss a great intellectual adventure. You will miss an exhilarating glimpse into one of the most marvelous, most useful creations of those small and mysterious computers inside our heads.

I am indebted to Dean Hickerson, Oliver Selfridge, and Peter Renz for looking over this book's manuscript and providing a raft of corrections and welcome suggestions.

—MARTIN GARDNER
January, 1998

WHAT IS A FUNCTION?

No concept in mathematics, especially in calculus, is more fundamental than the concept of a function. The term was first used in a 1673 letter written by Gottfried Wilhelm Leibniz, the German mathematician and philosopher who invented calculus independently of Isaac Newton. Since then the term has undergone a gradual extension of meaning.

In traditional calculus a function is defined as a relation between two terms called variables because their values vary. Call the terms x and y. If every value of x is associated with exactly one value of y, then y is said to be a function of x. It is customary to use x for what is called the *independent variable,* and y for what is called the *dependent variable* because its value depends on the value of x.

As Thompson explains in Chapter 3, letters at the end of the alphabet are traditionally applied to variables, and letters elsewhere in the alphabet (usually first letters such as $a,b,c. . .$) are applied to constants. Constants are terms in an equation that have a fixed value. For example, in $y = ax + b$, the variables are x and y, and a and b are constants. If $y = 2x + 7$, the constants are 2 and 7. They remain the same as x and y vary.

A simple instance of a geometrical function is the dependence of a square's area on the length of its side. In this case the function is called a one-to-one function because the dependency goes both ways. A square's side is also a function of its area.

A square's area is the length of its side multiplied by itself. To express the area as a function of the side, let y be the area, x the side, then write $y = x^2$. It is assumed, of course, that x and y are positive.

A slightly more complicated example of a one-to-one function is the relation of a square's side to its diagonal. A square's diagonal is the hypotenuse of an isosceles right triangle. We know from the Pythagorean theorem that the square of the hypotenuse equals the sum of the squares of the other two sides. In this case the sides are equal. To express the diagonal as a function of the square's side, let y be the diagonal, x the side, and write $y = \sqrt{2x^2}$, or more simply $y = x\sqrt{2}$ to express the side as a function of the diagonal, let y be the side, x the diagonal, and write $y = \sqrt{\dfrac{x^2}{2}}$, or more simply $y = \dfrac{x}{\sqrt{2}}$.

The most common way to denote a function is to replace y, the dependent variable, by $f(x)$—f being the first letter of "function." Thus $y = f(x) = x^2$ means that y, the dependent variable, is the square of x. Instead of, say, $y = 2x - 7$, we write $y = f(x) = 2x - 7$. This means that y, a function of x, depends on the value of x in the expression $2x - 7$. In this form the expression is called an *explicit* function of x. If the equation has the equivalent form of $2x - y - 7 = 0$, it is called an *implicit* function of x because the explicit form is implied by the equation. It is easily obtained from the equation by rearranging terms. Instead of $f(x)$, other symbols are often used.

If we wish to give numerical values to x and y in the example $y = f(x) = 2x - 7$, we replace x by any value, say 6, and write $y = f(6) = (2 \cdot 6) - 7$, giving the dependent variable y a value of 5.

If the dependent variable is a function of a single independent variable, the function is called a function of one variable. Familiar examples, all one-to-one functions, are:

The circumference or area of a circle in relation to its radius.

The surface or volume of a sphere in relation to its radius.

The log of a number in relation to the number.

Sines, cosines, tangents, and secants are called trigonometric functions. Logs are logarithmic functions. Exponential functions are functions in which x, the independent variable, is an exponent in a equation, such as $y = 2^x$. There are, of course, endless other examples of more complicated one-variable functions which have been given names.

Functions can depend on more than one variable. Again, there are endless examples. The hypotenuse of a right triangle depends on its two sides, not necessarily equal. (The function of course involves three variables, but it is called a two-variable function because it has two independent variables.) If z is the hypotenuse, we know from the Pythagorean theorem that $z = \sqrt{x^2 + y^2}$. Note that this is not a one-to-one function. Knowing x and y gives z a unique value, but knowing z does not yield unique values for x and y.

Two other familiar examples of a two-variable function, neither one-to-one, are the area of a triangle as a function of its altitude and base, and the area of a right circular cylinder as a function of its radius and height.

Functions of one and two variables are ubiquitous in physics. The period of a pendulum is a function of its length. The distance covered by a dropped stone and its velocity are each functions of the elapsed time since it was dropped. Atmospheric pressure is a function of altitude. A bullet's energy is a two-variable function dependent on its mass and velocity. The electrical resistance of a wire depends on the length of the wire and the diameter of its circular cross section.

Functions can have any number of independent variables. A simple instance of a three-variable function is the volume of a rectangular room. It is dependent on the room's two sides and height. The volume of a four-dimensional hyper-room is a function of four variables.

A beginning student of calculus must be familiar with how equations with two variables can be modeled by curves on the Cartesian plane. (The plane is named after the French mathematician and philosopher René Descartes who invented it.) Values of the independent variable are represented by points along the horizontal x axis. Values of the dependent variable are represented by points along the vertical y axis. Points on the plane signify an ordered pair of x and y numbers. If a function is linear— that is, if it has one form $y = ax + b$—the curve representing the ordered pairs is a straight line. If the function does not have the form $ax + b$ the curve is not a straight line.

Figure 1 is a Cartesian graph of $y = x^2$. The curve is a parabola. Points along each axis represent real numbers (rational and irra-

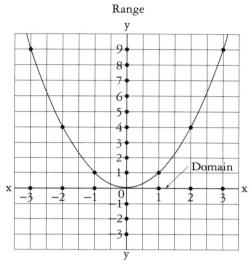

FIG. 1. $y = x^2$ or $f(x) = x^2$
Note that the scales are different on the two axes.

tional), positive on the right side of the x axis, negative on the left; positive at the top of the y axis, negative at the bottom. The graph's origin point, where the axes intersect, represents zero. If x is the side of a square, we assume it is neither zero nor negative, so the relevant curve would be only the right side of the parabola. Assume the square's side is 3. Move vertically up from 3 on the x axis to the curve, then go left to the y axis where you find that the square of 3 is 9. (I apologize to readers for whom all this is old hat.)

If a function involves three independent variables, the Cartesian graph must be extended to a three-dimensional space with axes x, y, and z. I once heard about a professor, whose name I no longer recall, who liked to dramatize this space to his students by running back and forth while he exclaimed "This is the x axis!" He then ran up and down the center aisle shouting "This is the y axis!", and finally hopped up and down while shouting "This is the z axis!" Functions of more than three variables require a Cartesian space with more than three axes. Unfortunately, a professor cannot dramatize axes higher than three by running or jumping.

Note the labels "domain" and "range" in Figure 1. In recent decades it has become fashionable to generalize the definition of function. Values that can be taken by the independent variable are called the variable's *domain.* Values that can be taken by the dependent variable are called the *range.* On the Cartesian plane the domain consists of numbers along the horizontal (x) axis. The range consists of numbers along the vertical (y) axis.

Domains and ranges can be infinite sets, such as the set of real numbers, or the set of integers; or either one can be a finite set such as a portion of real numbers. The numbers on a thermometer, for instance, represent a finite interval of real numbers. If used to measure the temperature of water, the numbers represent an interval between the temperatures at which water freezes and boils. Here the height of the mercury column relative to the water's temperature is a one-to-one function of one variable.

In modern set theory this way of defining a function can be extended to completely arbitrary sets of numbers for a function that is described not by an equation but by a set of rules. The simplest way to specify the rules is by a table. For example, the table in Figure 2 shows a set of arbitrary numbers that constitute the domain on the left. The corresponding set of arbitrary numbers in the range is on the right. The rules that govern this function are indicated by arrows. These arrows show that every number in the domain correlates to a single number on the right. As you can see, more than one number on the left can lead to the same number on the right, but not vice versa. Another example of such a function is shown in Figure 3, along with its graph, consisting of 6 isolated points in the plane.

Because every number on the left leads to exactly one number on the right, we can say that the numbers on the right are a function of those on the left. Some writers call the numbers on the right "images" of those on the left. The arrows are said to furnish a "mapping" of domain to range. Some call the arrows "correspondence rules" that define the function.

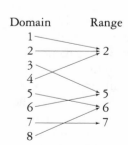

FIG. 2. An arbitrary function.

For most of the functions encountered in calculus, the domain consists of a single in-

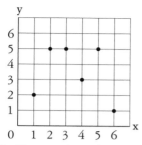 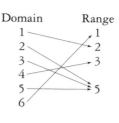

FIG. 3. How another arbitrary discrete function of integers is graphed.

terval of real numbers. The domain might be the entire x axis, as it is for the function $y = x^2$. Or it might be an interval that's bounded; for example, the domain of $y = \arcsin x$ consists of all x such that $-1 \leq x \leq 1$. Or it might be bounded on one side and unbounded on the other; for example, the domain of $y = \sqrt{x}$ consists of all $x \geq 0$. We call such a function "continuous" if its graph can be drawn without lifting the pencil from the paper, and "discontinuous" otherwise. (The complete definition of continuity, which is also applicable to functions with more complicated domains, is beyond the scope of this book.)

For example, the three functions just mentioned are all continuous. Figure 4 shows an example of a discontinuous function. Its domain consists of all real numbers, but its graph has infinitely many pieces that aren't connected to each other. In this book we will be concerned almost entirely with continuous functions.

Note that if a vertical line from the x axis intersects more than one point on a curve, the curve cannot represent a function because it maps an x number to more than one y number. Figure 5 is a graph that clearly is not a function because vertical lines, such as the one shown dotted, intersect the graph at three spots. (It should be noted that Thompson did not use the modern definition of "function." For example the graph shown in Figure 30 of Chapter XI fails this vertical line test, but Thompson considers it a function.)

In this generalized definition of function, a one-variable function is any set of ordered pairs of numbers such that every num-

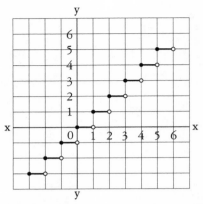

FIG. 4. This function is called the greatest integer function because it maps each real number (on the *x* axis) to the largest integer on the *y* axis that is equal to or less than the real number.

ber in one set is paired with exactly one number of the other set. Put differently, in the ordered pairs no *x* number can be repeated though a *y* number can be.

In this broad way of viewing functions, the arbitrary combination of a safe or the sequence of buttons to be pushed to open a door, are functions of counting numbers. To open a safe you must turn the knob back and forth to a random set of integers. If the safe's combination is, say, 2-19-3-2-19, then those numbers are a function of 1,2,3,4,5. They represent the order in which numbers must be taken to open the safe, or the order in which buttons must be pushed to open a door. In a similar way the heights of the tiny "peaks" along a cylinder lock's key are an arbitrary function of positions along the key's length.

In recent years mathematicians have widened the notion of function even further to include things that are not numbers. Indeed, they can be anything at all that are elements of a set. A function is simply the correlation of each el-

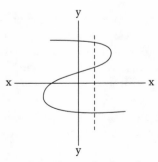

FIG. 5. A graph that does not represent a function.

ement in one set to exactly one element of another set. This leads to all sorts of uses of the word function that seem absurd. If Smith has red hair, Jones has black hair, and Robinson's hair is white, the hair color is a function of the three men. Positions of towns on a map are a function of their positions on the earth. The number of toes in a normal family is a function of the number of persons in the family. Different persons can have the same mother, but no person has more than one mother. This allows one to say that mothers are a function of persons. Elephant mothers are a function of elephants, but not grandmothers because an elephant can have two grandmothers. As one mathematician recently put it, functions have been generalized "up to the sky and down into the ground."

A useful way to think of functions in this generalized way is to imagine a black box with input and output openings. Any element in a domain, numbers or otherwise, is put into the box. Out will pop a single element in the range. The machinery inside the box magically provides the correlations by using whatever correspondence rules govern the function. In calculus the inputs and outputs are almost always real numbers, and the machinery in the black box operates on rules provided by equations.

Because the generalized definition of a function leads to bizarre extremes, many educators today, especially those with engineering backgrounds, think it is confusing and unnecessary to introduce such a broad definition of functions to beginning calculus students. Nevertheless, an increasing number of modern calculus textbooks spend many pages on the generalized definition. Their authors believe that defining a function as a mapping of elements from any set to any other set is a strong unifying concept that should be taught to all calculus students.

Opponents of this practice think that calculus should not be concerned with toes, towns, mothers, and elephants. Its domains and ranges should be confined, as they have always been, to real numbers whose functions describe continuous change.

It is a fortunate and astonishing fact that the fundamental laws of our fantastic fidgety universe are based on relatively simple equations. If it were otherwise, we surely would know less than we know now about how our universe behaves, and Newton and Leibniz would probably never have invented (or discovered?) calculus.

WHAT IS A LIMIT?

It is possible, though difficult, to understand calculus without a firm grasp on the meaning of a limit. A derivative, the fundamental concept of differential calculus, is a limit. An integral, the fundamental concept of integral calculus, is a limit.

To explain what is meant by a limit, we will be concerned in this chapter only with limits of discrete functions because limits are easier to understand in discrete terms. When you read *Calculus Made Easy* you will learn how the limit concept applies to what are called functions of a continuous variable because their variables have real number values that vary continuously. Functions of discrete variables have variables whose values jump from one value to another. There are also functions of complex variables in which the values are complex numbers—numbers based on the imaginary square root of minus one. Complex variables are outside the scope of Thompson's book.

A sequence is a set of numbers in some order. The numbers don't have to be different and they need not be integers. Consider the sequence 1,2,3,4,. . . . This is just the positive integers. It is an infinite sequence because it continues without stopping. If it stopped it would be a finite sequence.

If the terms of a finite sequence are added to obtain a finite sum, it is called a series. If a series is infinite, the sum up to any specified term is called a "partial sum." If the partial sums of an infinite series get closer and closer to a number k, so that by continuing the series you can make the sum as close to k as you please, then k is called the limit of the partial sums, or the limit of the infinite series. The terms are said to "converge" on k. If there is no convergence, the series is said to "diverge."

The limit of an infinite series is sometimes called its "sum at infinity," but of course this is not a sum in the usual arithmetical sense when the number of terms is finite. You can't obtain the "sum" of an infinite series by adding because the number of terms to be added is infinite. When we speak of the "sum" of an infinite series, this is just a short way of naming its limit.

An infinite series can converge on its limit in three different ways:

1. The partial sums get ever closer to the limit without actually reaching it, but they never go beyond the limit.
2. The partial sums reach the limit.
3. The partial sums go beyond the limit before they converge.

Let's look at examples of types 1 and 3.

The fifth century B.C. Greek philosopher Zeno of Elia invented several famous paradoxes intended to show that there is something extremely mysterious about motion. One of them imagines a runner going from A to B. He first runs half the distance, then half the remaining distance, then again half the remaining distance, and so on. The distances he runs get smaller and smaller in the halving series $\frac{1}{2} + \frac{1}{4} + \frac{1}{8} + \frac{1}{16} + \ldots \frac{1}{2^n}$. Distances from B approach zero as their limit while the distances from A form a series that converges on 1. The runner, of course, models a point moving along a line from A to B. Does the runner ever reach the goal?

It depends.

Assume that after each step in the series the runner pauses to rest for a second. We can model this with a pawn (representing a point) that you push across a table from one edge to the edge opposite. First you push the pawn half the distance, then pause for a second. You push it half the remaining distance and again pause for a second. If this procedure continues, the pawn (point) will get closer and closer to the limit, but will never reach it.

There is an old joke based on this. A mathematics professor places a male student at one side of an empty room and a gorgeous female student at the opposite wall. On command, the boy walks half the distance toward the girl, waits a second, then goes half the rest of the way, and so on, always pausing a second before he cuts the remaining distance in half. The girl says, "Ha ha, you'll never reach me!" The boy replies, "True, but I can get close enough for all practical purposes."

Suppose, now, that instead of waiting a second after each pawn push, the pawn is moved at a steady rate. Assume that the constant speed is such that the pawn goes half the distance in one second, half the remaining distance in half a second, and so on. No pauses. A discrete process has been transformed into a continuous one. In two seconds the pawn has reached the table's far edge. Zeno's runner, if he goes at a steady rate, will reach the goal in a finite period of time. The halving series, modeled in this fashion, converges exactly on the limit.

Zeno's runner leads to a variety of amusing paradoxes involving what are called "infinity machines." A simple example is a lamp that is turned off at the end of one minute, then turned on at the end of half a minute, off after a quarter minute, and so on in an infinite series of ons and offs. The time series converges on two minutes. At the end of two minutes is the lamp on or off? This of course is a thought experiment. It can't be performed with an actual lamp, but can it be answered in the abstract? No, because there *is* no last operation in an infinite series of on and off. It is like asking if the last digit of pi is odd or even.*

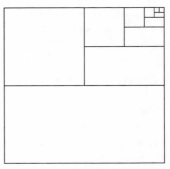

FIG. 6. A two-dimensional "look-see" proof that $\frac{1}{2} + \frac{1}{4} + \frac{1}{8} + \frac{1}{16} + \ldots = 1$.

An easy way to "see" that the limit of $\frac{1}{2} + \frac{1}{4} + \frac{1}{8} + \ldots$ is 1 is to mark off the fractional lengths along a number line as Thompson does in his Figure 46. A similar "look-see" proof that the series converges on 1 is shown by the dissected unit square in Figure 6. The

---

*On infinity machines, see "Alephs and Supertasks," Chapter 4, in my *Wheels, Life, and Other Mathematical Amusements* (W. H. Freeman, 1983), and the references cited in that chapter's bibliography.—M.G.

partial sums of this series are generated by the discrete function $1 - \frac{1}{2^n}$, where n takes the integral values $1,2,3,4,5, \ldots$.

We turn now to an infinite series that goes past its limit before finally converging. An example is provided by changing every other sign in the halving series to a minus sign: $\frac{1}{2} - \frac{1}{4} + \frac{1}{8} - \frac{1}{16} + \ldots$. The partial sums of this "alternating series" are alternately above and below the limit of $\frac{1}{3}$. The difference from $\frac{1}{3}$ can be made as small as you please, but every other partial sum is larger than the limit.

As an infinite series approaches but never reaches its limit, the differences between a partial sum and the limit get closer and closer to zero. Indeed they get so close that you can assume they *are* zero and therefore, as Thompson likes to say, they can be "thrown away." In early books on calculus, *terms* said to become infinitely close to zero were called "infinitesimals." Clearly there is something spooky about numbers living in a neverland that is infinitely close to zero, yet somehow not zero. In the halving series, for example, the fractions approaching zero never become infinitesimals because they always remain a finite portion of 1. Infinitesimals are an infinitely small part of 1. They are smaller than any finite fraction you can name, yet never zero. Are they legitimate mathematical entities, or should they be banished from mathematics?

The most outspoken opponent of infinitesimals was the eighteenth-century British philosopher Bishop George Berkeley who attacked them in a 1734 book titled *The Analyst, Or a Discourse Addressed To an Infidel Mathematician.* The infidel was the astronomer Edmond Halley, for whom Halley's comet is named, and the man who persuaded Newton to publish his famous *Principia.*

Here are some of Bishop Berkeley's complaints about infinitesimals. ("Fluxion" was Newton's term for a derivative.)

> And what are these fluxions? The velocities of evanescent increments. And what are these same evanescent increments? They are neither finite quantities, nor quantities infinitely small, nor yet nothing. May we not call them ghosts of departed quantities?

———

And of the aforesaid fluxions there be other fluxions, which fluxions of fluxions are called second fluxions. And the fluxions of these second fluxions are called third fluxions: and so on, fourth, fifth, sixth, etc., *ad infinitum.* Now, as our Sense is strained and puzzled with the perception of objects extremely minute, even so the Imagination, which faculty derives from sense, is very much strained and puzzled to frame clear ideas of the least particle of time, or the least increment generated therein: and much more to comprehend the moments, or those increments of the flowing quantities in *status nascenti,* in their first origin or beginning to exist, before they become finite particles. And it seems still more difficult to conceive the abstracted velocities of such nascent imperfect entities. But the velocities of the velocities, the second, third, fourth, and fifth velocities, etc., exceed, if I mistake not, all human understanding. The further the mind analyseth and pursueth these fugitive ideas the more it is lost and bewildered; the objects, at first fleeting and minute, soon vanishing out of sight. Certainly, in any sense, a second or third fluxion seems an obscure Mystery. The incipient celerity of an incipient celerity, the nascent augment of a nascent augment, i. e. of a thing which hath no magnitude; take it in what light you please, the clear conception of it will, if I mistake not, be found impossible; whether it be so or no I appeal to the trial of every thinking reader. And if a second fluxion be inconceivable, what are we to think of third, fourth, fifth fluxions, and so on without end.

He who can digest a second or third fluxion, a second or third difference, need not, methinks, be squeamish about any point in Divinity.

Johann Bernoulli, a Swiss mathematician who did pioneering work in developing calculus, expressed the paradox of infinitesimals crisply. They are so tiny, he said, that "if a quantity is increased or decreased by an infinitesimal, then that quantity is neither increased nor decreased."

For two centuries most mathematicians agreed with Berkeley and refused to use the term. You won't find it in *Calculus Made*

Easy. Bertrand Russell, in *Principles of Mathematics* (1903, Chapters 39 and 40) has a vigorous attack on infinitesimals. He calls them "mathematically useless," "unnecessary, erroneous, and self-contradictory." As late as 1941 the noted mathematician Richard Courant wrote: "Infinitely small quantities are now definitely and dishonorably discarded." Like Russell and others, he believed that calculus should replace infinitesimals by the concept of limits.

Charles Peirce (1839-1914), America's great mathematician and philosopher, and friend of William James, strongly disagreed. He was almost alone in his day in siding with Leibniz, who believed that infinitesimals were as real and as legitimate as imaginary numbers. Here are some typical remarks by Peirce that I found by checking "infinitesimal" in the indexes of the volumes that make up Peirce's *Collected Papers* and his *New Elements of Mathematics.*

Infinitesimals may exist and be highly important for philosophy, as I believe they are.

The doctrine of infinitesimals is far simpler than the doctrine of limits.

Is it consistent . . . freely to admit of imaginaries while rejecting infinitesimals as inconceivable?

Infinitesimals, in the strict and literal sense, are perfectly intelligible, contrary to the teaching of the great body of modern textbooks on the calculus.

There is nothing contradictory about the idea of such quantities. . . . As a mathematician, I prefer the method of infinitesimals to that of limits, as far easier and less infested with snares.

Peirce would have been delighted had he lived to see the work of Abraham Robinson, of Yale University. In 1960, to the vast surprise of mathematicians everywhere, Robinson found a way to reintroduce Leibniz's infinitesimals as legitimate, precisely defined mathematical entities! His way of using them in calculus is known as "nonstandard analysis." (Analysis is a term applied to calculus and all higher mathematics that use calculus.) Nonstandard analysis has produced simpler solutions than standard analysis to many calculus problems, and of course it is closer to an intuitive way of interpreting infinite converging series. Robinson's achievement is too difficult to go into here, but you will find a good introduction to it in "Nonstandard Analysis," by Martin Davis and Reuben Hersh, in *Scientific American,* June 1972.

Mathematician and science fiction writer Rudy Rucker, in his book *Infinity and the Mind* (1982) vigorously defends infinitesimals:

> So great is the average person's fear of infinity that to this day calculus all over the world is being taught as a study of *limit processes* instead of what it really is: *infinitesimal analysis.*
>
> As someone who has spent a good portion of his adult life teaching calculus courses for a living, I can tell you how weary one gets of trying to explain the complex and fiddling theory of limits to wave after wave of uncomprehending freshmen. . . .
>
> But there is hope for a brighter future. Robinson's investigations of the hyperreal numbers have put infinitesimals on a logically unimpeachable basis, and here and there calculus texts based on infinitesimals have appeared.

Which is preferable? To talk about quantities so infinitely small that you can, as Thompson says, "throw them away," or to talk of values approaching a limit? Debate over the infinitesimal versus the limit language goes nowhere because they are two ways of saying the same thing. It's like choosing between calling a triangle a polygon with three sides or a polygon with three angles. Calculations in differentiating or integrating are exactly the same regardless of your preference for how to describe what you are doing. Now that infinitesimals have become respectable again, thanks to nonstandard analysis, you needn't hesitate, if you like, to use the term.

You might suppose that if the terms of an infinite series get smaller and smaller, the series must converge. This is far from true. The most famous example is $\frac{1}{1} + \frac{1}{2} + \frac{1}{3} + \frac{1}{4} + \frac{1}{5} + \dots$. Known as the "harmonic series," it has countless applications in physics as well as in mathematics. Although its fractions get progressively smaller, converging on zero, its partial sums grow without limit! The growth is infuriatingly slow. After a hundred terms the partial sum is only a bit higher than 5. To reach a sum of 100 requires more than 10^{43} terms!

If we eliminate all terms in the harmonic series that have even denominators, will it converge? Amazingly, it will not, though its rate of growth is much slower. If we eliminate from the series all terms whose denominators contain a specific digit one or more times, the series *will* then converge. The following table gives to two decimal places the limit for each omitted digit:

Omitted digit	Sum
1	16.18
2	19.26
3	20.57
4	21.33
5	21.83
6	22.21
7	22.49
8	22.73
9	22.92
0	23.10

Limits of infinite series can be expressed by unending decimal fractions. For example, .33333. . . . is the limit of the series $\frac{3}{10} + \frac{3}{100} + \frac{3}{1000} + \ldots = \frac{1}{3}$. Incidentally, there is a ridiculously easy way to determine the integral limit of any repeating decimal. The trick is to divide the repetend (the repeated sequence of digits) by a number consisting of the same number of nines as there are digits in the repetend. Thus .3333. . . reduces to $\frac{3}{9} = \frac{1}{3}$. If the repeating decimal is, say, .123123123. . . . the limit is $\frac{123}{999}$ which reduces to $\frac{41}{333}$.

Irrational numbers such as irrational roots, and transcendental numbers such as pi and e, are limits of many infinite series. Pi, for example, is the limit of such highly patterned series as: $\frac{4}{1} - \frac{4}{3} + \frac{4}{5} - \frac{4}{7} + \frac{4}{9} - \ldots$. The number e (you will encounter it in Thompson's Chapter 14) is the limit of $1 + \frac{1}{1!} + \frac{1}{2!} + \frac{1}{3!} + \frac{1}{4!} + \ldots$.

Although Archimedes did not know calculus, he anticipated integration by calculating pi as the limit of the perimeters of regular polygons as their number of sides increases. In the language of infinitesimals, a circle can be viewed as the perimeter of a regular polygon with an infinity of sides, its perimeter consisting of an infinity of straight line segments each of infinitesimal length.

Many ingenious techniques have been found for determining if an infinite series converges or diverges, as well as ways, sometimes not easy, of finding the limit. If the terms of a series decrease in a geometric progression (each term is the same fraction of the preceding one) finding the limit is easy. Here is how it works on the halving series $1 + \frac{1}{2} + \frac{1}{4} + \frac{1}{8} + \ldots$. Let x equal the entire series. Multiply each side of the equation by 2:

$$2x = 2 + \frac{2}{2} + \frac{2}{4} + \frac{2}{8} + \frac{2}{16} + \ldots$$

Reduce the terms:

$$2x = 2 + 1 + \frac{1}{2} + \frac{1}{4} + \frac{1}{8} + \ldots$$

Note that the series beyond 2 is the same as the original halving series which we took as x. This enables us to substitute x for the sequence and write $2x = 2 + x$. Rearranging terms to $2x - x = 2$ gives x, the limit of the series, a value of 2.

The same trick will show that $\frac{1}{2}$ is the limit of $\frac{1}{3} + \frac{1}{9} + \frac{1}{27} + \frac{1}{81} + \ldots$; it works on any series in which terms decrease in geometric progression.

Bouncing ball problems are common in the literature on limits. They assume that an ideally elastic ball is dropped a specified distance to a hard floor. After each bounce it rises a constant fraction of the previous height. Here is a typical example.

The ball is dropped from a height of four feet. Each bounce takes it to $\frac{3}{4}$ the previous height. In practice, of course, a rubber ball bounces only a finite number of times, but the idealized ball bounces an infinite number of times. The rises approach zero as a limit, but because the times of each bounce also approach a limit of zero, the ball (like Zeno's runner) finally reaches the limit. After an infinity of bounces, it comes to rest after a finite period of time. When the ball ceases to bounce, how far has it traveled?

We can solve this problem by using the same trick used on the halving series. Ignoring for a moment the initial drop of four feet, the ball will rise three feet then fall three feet for a total of six feet. After that, each bounce (rise plus fall) is three-fourths the previous bounce. Letting x be the total distance the ball travels after the first drop of four feet, we write the equation:

$$x = 6 + \frac{18}{4} + \frac{54}{16} + \frac{162}{64} + \frac{486}{256} + \dots$$

Reducing the fractions:

$$x = 6 + \frac{9}{2} + \frac{27}{8} + \frac{81}{32} + \frac{243}{128} + \dots$$

Because each term is $\frac{4}{3}$ its following term, we multiply each side by $\frac{4}{3}$ to get:

$$\frac{4x}{3} = 8 + 6 + \frac{9}{2} + \frac{27}{8} + \frac{81}{32} + \dots .$$

Observe that after 8 the sequence is the same as x, so we can substitute x for it:

$$\frac{4x}{3} = 8 + x$$

$$4x = 24 + 3x$$

$$x = 24$$

This is the distance the ball bounces after the initial drop of 4 feet. The total distance traveled by the ball is $24 + 4 = 28$ feet.

Sam Loyd, America's great puzzlemaker, in his *Cyclopedia of Puzzles* (page 23), and his British counterpart Henry Ernest Dudeney, in *Puzzles and Curious Problems* (Problem 223), each give the following ball bounce problem. A ball is dropped 179 feet from the Tower of Pisa. Each bounce is one-tenth the height of the previous bounce. How far does the ball travel after an infinity of bounces before it finally comes to rest? (See Figure 7.)

We can solve this by the trick used before, but because each fraction is one-tenth the previous one, we can find the answer by an even faster method.

After the initial drop of 179 feet, the height of the first bounce is 17.9. Succeeding bounces have heights of 1.79, .179, .0179, and so on, with the decimal point moving one position left after each bounce. Adding these heights gives a total of 19.8888. . . . We now double this distance to obtain the up plus down distances for each bounce to get 39.7777. . . . Finally, we add the

THE LEANING TOWER ~OF~ PISA
A classical puzzle BY SAM LOYD.

FIG. 7. Sam Loyd's bouncing ball puzzle.

initial drop of 179 feet to obtain the total distance the ball travels: 218.7777..., or exactly 218 and $\frac{7}{9}$ feet.

Converging series that do not decrease by a geometric progression often can be solved by other clever methods. Here is an interesting example.

$$x = 1 + \tfrac{3}{2} + \tfrac{5}{4} + \tfrac{7}{8} + \tfrac{9}{16} + \tfrac{11}{32} + \dots.$$

Note that the numerators are odd numbers in sequence, and the denominators are a doubling series. Here is a simple way to find the limit.

First divide each term by 2:

$$\tfrac{x}{2} = \tfrac{1}{2} + \tfrac{3}{4} + \tfrac{5}{8} + \tfrac{7}{16} + \dots.$$

Subtract this sequence from the original sequence:

$$x = 1 + \tfrac{3}{2} + \tfrac{5}{4} + \tfrac{7}{8} + \tfrac{9}{16} + \dots$$
$$\tfrac{x}{2} = \quad\ \ \tfrac{1}{2} + \tfrac{3}{4} + \tfrac{5}{8} + \tfrac{7}{16} + \dots$$
$$\overline{\tfrac{x}{2} = 1 + \left[1 + \tfrac{1}{2} + \tfrac{1}{4} + \tfrac{1}{8} + \dots.\right]}$$

Observe that after the 1 inside the brackets, the sequence that follows is our old friend the halving series which we know converges on 1. Adding 2 to the initial 1 gives the series a limit of 3. Since 3 is half of x, x must be 6, the limit of the original series.

Thompson does not spend much time on series and their limits. I have done so in this chapter for two reasons: they are the best way to become comfortable with the limit concept, and modern calculus textbooks now usually include chapters on infinite series and their usefulness in many aspects of calculus.

WHAT IS A DERIVATIVE?

In Chapter 3 Thompson makes crystal clear what a derivative is, and how to calculate it. However, it seemed to me useful to make a few introductory remarks about derivatives that may make Thompson's chapter even easier to understand.

Let's start with Zeno's runner. Assume that he runs ten meters per second on a path from zero to 100 meters. The independent variable is time, represented by the x axis of a Cartesian graph. The dependent variable y is the runner's distance from his starting spot. It is represented on the y axis. Because the function is linear, the runner's motion graphs as an upward tilted straight line from zero, the graph's origin, to the point that is ten seconds on the time axis and 100 meters on the distance axis. (Figure 8) If by distance we mean distance *from* the goal, the line on the graph tilts the other way (Figure 9).

Given any point in time, how fast is the runner moving? Because we are dealing with a simple linear function we don't need calculus to tell us that at every instant he is going ten meters per second. The function's equation is $y = 10x$. Note that the slope of the line on the graph, as measured by the height in meters at any point divided by the elapsed time in seconds at any point, is 10. At each instant the runner has gone in meters ten times the elapsed number of seconds. His instantaneous speed throughout the run clearly is ten meters per second.

Consider any moment of time along the x axis, then go vertically up the graph to the distance traveled in meters. You will find that the distance is always ten times the elapsed time. As you will learn when you read *Calculus Made Easy,* the derivative of a

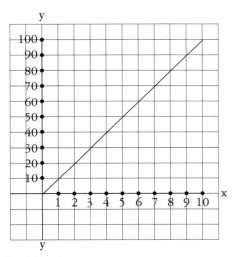

FIG. 8. Graph of Zeno's runner.
The *x* axis is time, the *y* axis is the distance from the start of the run.

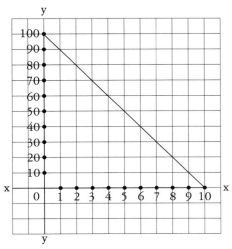

FIG. 9. Graph of Zeno's runner showing distance from goal. The equation is $y = 10(10 - x)$.

function is simply another function that describes the rate at which a dependent variable changes with respect to the rate at which the independent variable changes. In this case the runner's speed never changes, so the derivative of $y = 10x$ is simply the number 10. It tells you two things: (1) that at any time the runner's speed is ten meters per second, and (2) that at any point on the line that graphs this function, the slope of the line is 10. This generalizes to all linear functions in which the variable y changes with respect to variable x at a constant rate. If a function is $y = ax$, its derivative is simply the constant a.

As I said, you don't need calculus to tell you all this, but it is good to know that calculating derivatives gives the correct result even when functions are linear.

An even simpler case of a derivative, too obvious to require any thought, let alone demanding calculus, is the case of a runner who stands perfectly still. Let's say he stops running after going ten meters. The function is $y = 10$. The graph becomes a horizontal straight line as shown in Figure 10. Its slope is zero which is the same as saying that the rate at which the runner's distance from the start changes, relative to changes of time, is zero. The function's derivative is zero. Even in this extreme case it is comforting to know that calculus still applies. In general, the derivative of any constant is zero.

Calculus ceases to be trivial when functions are nonlinear. Consider the simple nonlinear function $y = x^2$, which Thompson uses to open his chapter on derivatives. Let's see how it applies to the growth of a square, the simplest geometrical interpretation of this function.

Imagine a monster living on Flatland, a plane of two dimensions. It is born a square of side 1 and area 1, then grows at a steady rate. We wish to know, at any instant of time, how fast its area grows with respect to the growth of its side.

The monster's area, of course, is the square of its side, so the function we have to consider is

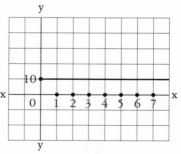

FIG. 10. Graph of a runner who stands still at a distance of ten units from the start.

$y = x^2$, where y is the area and x the side. (It graphs as the parabola shown in Figure 1 of the first preliminary chapter.) As you will learn from Thompson, the function's derivative is $2x$. What does this tell us? It tells us that at any given moment the monster's area is growing at a rate that is $2x$ times as fast as its side is lengthening.

For example, let's say the monster's side is growing at a rate of 3 units per second. Starting with a side of one unit, at the end of ten seconds its side will have reached 31 units. The value of x at this point is 31. The derivative says that when the monster's side is 31, its area is increasing with respect to its side at a rate of $2x$, or $2 \times 31 = 62$ units. When the square reaches 100 on its side, its area will be increasing with respect to its side by $2 \times 100 = 200$ units.

These figures express the rate of the square's growth with respect to its side. For the square's growth rate with respect to *time* we have to multiply these values by 3. Thus when the square has a side of 31 (after ten seconds), it is growing at a rate of $3 \times 2 \times 31 = 186$ square units per second. When the side is 100, its growth rate per second is $3 \times 2 \times 100 = 600$.

Suppose the monster is a cube of edge x which increases at a steady rate of 2 units per second. The cube's volume, y, is x^3. The derivative of the function $y = x^3$ is $3x^2$. This tells you that the cube's volume in cubical units grows $3x^2$ times as fast as its edge grows. Thus when the cube's side reaches, say, 10, the value of x, its volume, is growing $3 \times 10^2 = 300$ square inches as fast as its side. Its growth rate per second is $2 \times 3 \times 10^2 = 600$.

Although Thompson avoids defining a derivative as the limit of a sequence of ratios, this clearly is the case. Suppose, for instance, that our growing square has sides that increase at one unit per second. We can tabulate the area's growth at times slightly larger than 2 seconds as follows:

Time	Side	Area
2	3	9
2.1	3.1	9.61
2.01	3.01	9.0601
2.001	3.001	9.006001

The average rate of growth from time 2 to time 2.1 is:

$$\frac{9.61 - 9}{2.1 - 2} = 6.1$$

And from time 2 to time 2.01:

$$\frac{9.0601 - 9}{2.01 - 2} = 6.01$$

And from time 2 to time 2.001:

$$\frac{9.006001 - 9}{2.001 - 2} = 6.001$$

The averages obviously approach a limit of 6. Thus the derivative of the area with respect to time is the limit of an infinite sequence of ratios that converge on 6. Put simply, a derivative is the rate at which a function's dependent variable grows with respect to the growth rate of the independent variable. In geometrical terms, it determines the exact slope of the tangent to a function's curve at any specified point along the curve. This equivalence of the algebraic and geometrical definitions of a derivative is one of the most beautiful aspects of calculus.

I hope this and the previous two preliminary chapters will help prepare you for understanding *Calculus Made Easy.*

CALCULUS MADE EASY

What one fool can do, another can.

—Ancient Simian proverb

PUBLISHER'S NOTE ON THE THIRD EDITION

Only once in its long and useful life in 1919, has this book been enlarged and revised. But in twenty-six years much progress can be made, and the methods of 1919 are not likely to be the same as those of 1945. If, therefore, any book is to maintain its usefulness, it is essential that it should be overhauled occasionally so that it may be brought up-to-date where possible, to keep pace with the forward march of scientific development.

For the new edition the book has been reset, and the diagrams modernised. Mr. F. G. W. Brown has been good enough to revise the whole of the book, but he has taken great care not to interfere with the original plan. Thus teachers and students will still recognise their well-known guide to the intricacies of the calculus. While the changes made are not of a major kind, yet their significance may not be inconsiderable. There seems no reason now, even if one ever existed, for excluding from the scope of the text those intensely practical functions, known as the hyperbolic sine, cosine and tangent, whose applications to the methods of integration are so potent and manifold. These have, accordingly, been introduced and applied, with the result that some of the long cumbersome methods of integrating have been displaced, just as a ray of sunshine dispels an obstructing cloud.

The introduction, too, of the very practical integrals:

$$\int e^{pt} \sin kt \cdot dt \quad \text{and} \quad \int e^{pt} \cos kt \cdot dt$$

has eliminated some of the more ancient methods of "Finding Solutions" (Chapter XXI). By their application, shorter and more intelligible ones have grown up naturally instead.

In the treatment of substitutions, the whole text has been tidied up in order to render it methodically consistent. A few examples have also been added where space permitted, while the whole of the exercises and their answers have been carefully revised, checked and corrected. Duplicated problems have thus been removed and many hints provided in the answers adapted to the newer and more modern methods introduced.

It must, however, be emphatically stated that the plan of the original author remains unchanged; even in its more modern form, the book still remains a monument to the skill and the courage of the late Professor Silvanus P. Thompson. All that the present reviser has attempted is to revitalize the usefulness of the work by adapting its distinctive utilitarian bias more closely in relation to present-day requirements.

PROLOGUE

Considering how many fools can calculate, it is surprising that it should be thought either a difficult or a tedious task for any other fool to learn how to master the same tricks.

Some calculus-tricks are quite easy. Some are enormously difficult. The fools who write the text-books of advanced mathematics—and they are mostly clever fools—seldom take the trouble to show you how easy the easy calculations are. On the contrary, they seem to desire to impress you with their tremendous cleverness by going about it in the most difficult way.

Being myself a remarkably stupid fellow, I have had to unteach myself the difficulties, and now beg to present to my fellow fools the parts that are not hard. Master these thoroughly, and the rest will follow. What one fool can do, another can.

COMMON GREEK LETTERS USED AS SYMBOLS

Capital	Small	English Name	Capital	Small	English Name
A	α	Alpha	Λ	λ	Lambda
B	β	Beta	M	μ	Mu
Γ	γ	Gamma	Ξ	ξ	Xi
Δ	δ	Delta	Π	π	Pi
E	ϵ	Epsilon	P	ρ	Rho
H	η	Eta	Σ	σ	Sigma
Θ	θ	Thēta	Φ	ϕ	Phi
K	κ	Kappa	Ω	ω	Omega

TO DELIVER YOU FROM THE PRELIMINARY TERRORS

———✵———

The preliminary terror, which chokes off most high school students from even attempting to learn how to calculate, can be abolished once for all by simply stating what is the meaning—in common-sense terms—of the two principal symbols that are used in calculating.

These dreadful symbols are:

(1) d which merely means "a little bit of".

Thus dx means a little bit of x; or du means a little bit of u. Ordinary mathematicians think it more polite to say "an element of", instead of "a little bit of". Just as you please. But you will find that these little bits (or elements) may be considered to be infinitely small.

(2) $\int$ which is merely a long S, and may be called (if you like) "the sum of".

Thus $\int dx$ means the sum of all the little bits of x; or $\int dt$ means the sum of all the little bits of t. Ordinary mathematicians call this symbol "the integral of". Now any fool can see that if x is considered as made up of a lot of little bits, each of which is called dx, if you add them all up together you get the sum of all the dx's (which is the same thing as the whole of x). The word "integral" simply means "the whole". If you think of the duration of time for one hour, you may (if you like) think of it as cut up into 3600 little bits called seconds.

The whole of the 3600 little bits added up together make one hour.

When you see an expression that begins with this terrifying symbol, you will henceforth know that it is put there merely to give you instructions that you are now to perform the operation (if you can) of totalling up all the little bits that are indicated by the symbols that follow.

That's all.

Chapter II

ON DIFFERENT DEGREES OF SMALLNESS

We shall find that in our processes of calculation we have to deal with small quantities of various degrees of smallness.

We shall have also to learn under what circumstances we may consider small quantities to be so minute that we may omit them from consideration. Everything depends upon relative minuteness.

Before we fix any rules let us think of some familiar cases. There are 60 minutes in the hour, 24 hours in the day, 7 days in the week. There are therefore 1440 minutes in the day and 10,080 minutes in the week.

Obviously 1 minute is a very small quantity of time compared with a whole week. Indeed, our forefathers considered it small as compared with an hour, and called it "one minute", meaning a minute fraction—namely one sixtieth—of an hour. When they came to require still smaller subdivisions of time, they divided each minute into 60 still smaller parts, which, in Queen Elizabeth's days, they called "second minutes" (*i.e.* small quantities of the second order of minuteness). Nowadays we call these small quantities of the second order of smallness "seconds". But few people know *why* they are so called.

Now if one minute is so small as compared with a whole day, how much smaller by comparison is one second!

Again, think of a hundred dollars compared with a penny: it is worth only a $\frac{1}{100}$ part. A penny is of precious little importance compared with a hundred dollars: it may certainly be regarded as a *small* quantity. But compare a penny with ten thousand dollars: relative to this greater sum, a penny is of no more importance than a hundredth of a penny would be to a hundred dollars. Even a hundred dollars is relatively a negligible quantity in the wealth of a millionaire.

Now if we fix upon any numerical fraction as constituting the proportion which for any purpose we call relatively small, we can easily state other fractions of a higher degree of smallness. Thus if, for the purpose of time, $\frac{1}{60}$ be called a *small* fraction, then $\frac{1}{60}$ of $\frac{1}{60}$ (being a *small* fraction of a *small* fraction) may be regarded as a *small quantity of the second order* of smallness.*

Or, if for any purpose we were to take 1 percent $\left(i.e.\ \frac{1}{100}\right)$ as a *small* fraction then 1 percent of 1 percent $\left(i.e.\ \frac{1}{10,000}\right)$ would be a small fraction of the second order of smallness; and $\frac{1}{1,000,000}$ would be a small fraction of the third order of smallness, being 1 percent of 1 percent of 1 percent.

Lastly, suppose that for some very precise purpose we should regard $\frac{1}{1,000,000}$ as "small". Thus, if a first-rate chronometer is not to lose or gain more than half a minute in a year, it must keep time with an accuracy of 1 part in 1,051,200. Now if, for such a purpose, we regard $\frac{1}{1,000,000}$ (or one millionth) as a small quantity, then $\frac{1}{1,000,000}$ of $\frac{1}{1,000,000}$, that is, $\frac{1}{1,000,000,000,000}$ will be a small quantity of the second order of smallness, and may be utterly disregarded, by comparison.

Then we see that the smaller a small quantity itself is, the more negligible does the corresponding small quantity of the second order become. Hence we know that *in all cases we are justified in neglecting the small quantities of the second—or third* (or higher)—*orders, if only we take the small quantity of the first order small enough in itself.*

But it must be remembered that small quantities, if they occur in our expressions as factors multiplied by some other factor, may become important if the other factor is itself large. Even a penny becomes important if only it is multiplied by a few hundred.

Now in the calculus we write dx for a little bit of x. These things such as dx, and du, and dy, are called "differentials", the differential of x, or of u, or of y, as the case may be. [You *read*

*The mathematicians may talk about the second order of "magnitude" (*i.e.* greatness) when they really mean second order of *smallness*. This is very confusing to beginners.

them as *dee-eks*, or *dee-you*, or *dee-wy*.] If dx be a small bit of x, and relatively small of itself, it does not follow that such quantities as $x \cdot dx$, or $x^2 dx$, or $a^x dx$ are negligible. But $dx \times dx$ would be negligible, being a small quantity of the second order.

A very simple example will serve as illustration. Consider the function $f(x) = x^2$

Let us think of x as a quantity that can grow by a small amount so as to become $x + dx$, where dx is the small increment added by growth. The square of this is $x^2 + 2x \cdot dx + (dx)^2$. The second term is not negligible because it is a first-order quantity; while the third term is of the second order of smallness, being a bit of a bit of x^2. Thus if we took dx to mean numerically, say, $\frac{1}{60}$ of x, then the second term would be $\frac{2}{60}$ of x^2, whereas the third term would be $\frac{1}{3,600}$ of x^2. This last term is clearly less important than the second. But if we go further and take dx to mean only $\frac{1}{1000}$ of x, then the second term will be $\frac{2}{1000}$ of x^2, while the third term will be only $\frac{1}{1,000,000}$ of x^2.

Geometrically this may be depicted as follows: Draw a square (Fig. 1) the side of which we will take to represent x. Now suppose the square to grow by having a bit dx added to its size each way. The enlarged square is made up of the original square x^2, the two rectangles at the top and on the right, each of which is of area $x \cdot dx$ (or together $2x \cdot dx$), and a little square at the top right-hand corner which is $(dx)^2$. In Fig. 2 we have taken dx as quite a big fraction of x—about $\frac{1}{5}$. But suppose we had taken it only $\frac{1}{100}$—about the thickness of an inked line drawn with a fine pen (See Figure 3). Then the little corner square will have an area of only $\frac{1}{10,000}$ of x^2, and be practically invisible. Clearly $(dx)^2$ is negligible if only we consider the increment dx to be itself small enough.

Let us consider a simile.

Suppose a millionaire were to say to his secretary: next week I will give you a small fraction of any money that comes in to me. Suppose that the secretary were to say to his boy: I will give you a small fraction of what I get. Suppose the fraction in each case to be $\frac{1}{100}$ part. Now if Mr. Millionaire received during the next week \$1,000, the secretary would receive \$10 and the boy 1 dime. Ten dollars would be a small quantity com-

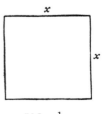

FIG. 1.

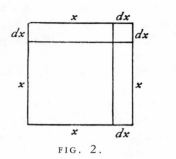

FIG. 2.

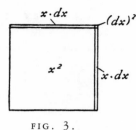

FIG. 3.

pared with $1,000; but a dime is a small small quantity indeed, of a very secondary order. But what would be the disproportion if the fraction, instead of being $\frac{1}{100}$, had been settled at $\frac{1}{1000}$ part? Then, while Mr. Millionaire got his $1,000, Mr. Secretary would get only $1.00, and the boy only a tenth of a penny!

The witty Dean Swift once wrote:

> So, Nat'ralists observe, a Flea
> Hath smaller Fleas that on him prey.
> And these have smaller Fleas to bite 'em.
> And so proceed *ad infinitum*.

An ox might worry about a flea of ordinary size—a small creature of the first order of smallness. But he would probably not trouble himself about a flea's flea; being of the second order of smallness, it would be negligible. Even a gross of fleas' fleas would not be of much account to the ox.

Chapter III

ON RELATIVE GROWINGS

All through the calculus we are dealing with quantities that are growing, and with rates of growth. We classify all quantities into two classes: *constants* and *variables*. Those which we regard as of fixed value, and call *constants,* we generally denote algebraically by letters from the beginning of the alphabet, such as a, b, or c; while those which we consider as capable of growing, or (as mathematicians say) of "varying", we denote by letters from the end of the alphabet, such as x, y, z, u, v, w, or sometimes t.

Moreover, we are usually dealing with more than one variable at once, and thinking of the way in which one variable depends on the other: for instance, we think of the way in which the height reached by a projectile depends on the time of attaining that height. Or, we are asked to consider a rectangle of given area, and to enquire how any increase in the length of it will compel a corresponding decrease in the breadth of it. Or, we think of the way in which any variation in the slope of a ladder will cause the height that it reaches, to vary.

Suppose we have got two such variables that depend on each other. An alteration in one will bring about an alteration in the other, *because* of this dependence. Let us call one of the variables x, and the other that depends on it y.

Suppose we make x to vary, that is to say, we either alter it or imagine it to be altered, by adding to it a bit which we call dx. We are thus causing x to become $x + dx$. Then, because x has been altered, y will have altered also, and will have become $y + dy$. Here the bit dy may be in some cases positive, in others negative; and it won't (except very rarely) be the same size as dx.

Take Two Examples.

(1) Let x and y be respectively the base and the height of a right-angled triangle (Fig. 4), of which the slope of the other side is fixed at 30°. If we suppose this triangle to expand and yet keep its angles the same as at first, then, when the base grows so as to become $x + dx$, the height be-

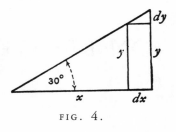

FIG. 4.

comes $y + dy$. Here, increasing x results in an increase of y. The little triangle, the height of which is dy, and the base of which is dx, is similar to the original triangle; and it is obvious that the

value of the ratio $\dfrac{dy}{dx}$ is the same as that of the ratio $\dfrac{y}{x}$. As the angle is 30° it will be seen that here[1]

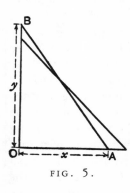

FIG. 5.

$$\frac{dy}{dx} = \frac{1}{1.73\ldots}.$$

(2) Let x represent, in Fig. 5, the horizontal distance, from a wall, of the bottom end of a ladder, AB, of fixed length; and let y be the height it reaches up the wall. Now y clearly depends on x. It is easy to see that, if we pull the bottom end A a bit farther from the wall, the top end B will come down a little lower. Let us state this in scientific language. If we increase x to $x + dx$, then y will become $y - dy$; that is, when x receives a positive increment, the increment which results to y is negative.

Yes, but how much? Suppose the ladder was so long that when the bottom end A was 19 inches from the wall the top end B reached just 15 feet from the ground. Now, if you were to pull the bottom end out 1 inch more, how much would the top end come down? Put it all into inches: $x = 19$ inches, $y = 180$ inches. Now the increment of x which we call dx, is 1 inch: or $x + dx = 20$ inches.

---

1. The cotangent of 30° is $\sqrt{3} = 1.7320\ldots$. Its reciprocal, $\frac{1}{1.7320}\ldots$, is .5773..., the tangent of 30°.—M.G.

How much will y be diminished? The new height will be $y - dy$. If we work out the height by the Pythagorean Theorem, then we shall be able to find how much dy will be. The length of the ladder is

$$\sqrt{(180)^2 + (19)^2} = 181 \text{ inches.}$$

Clearly, then, the new height, which is $y - dy$, will be such that

$$(y - dy)^2 = (181)^2 - (20)^2 = 32761 - 400 = 32361,$$
$$y - dy = \sqrt{32361} = 179.89 \text{ inches.}$$

Now y is 180, so that dy is $180 - 179.89 = 0.11$ inch.

So we see that making dx an increase of 1 inch has resulted in making dy a decrease of 0.11 inch.

And the ratio of dy to dx may be stated thus:

$$\frac{dy}{dx} = \frac{0.11}{1}$$

It is also easy to see that (except in one particular position) dy will be of a different size from dx.

Now right through the differential calculus we are hunting, hunting, hunting for a curious thing, a mere ratio, namely, the proportion which dy bears to dx when both of them are infinitely small.

It should be noted here that we can find this ratio $\dfrac{dy}{dx}$ only when y and x are related to each other in some way, so that whenever x varies y does vary also. For instance, in the first example just taken, if the base x of the triangle be made longer, the height y of the triangle becomes greater also, and in the second example, if the distance x of the foot of the ladder from the wall be made to increase, the height y reached by the ladder decreases in a corresponding manner, slowly at first, but more and more rapidly as x becomes greater. In these cases the relation between x and y is perfectly definite, it can be expressed mathematically, being $\dfrac{y}{x} = \tan 30°$ and $x^2 + y^2 = l^2$ (where l is the length of the ladder) respectively, and $\dfrac{dy}{dx}$ has the meaning we found in each case.

If, while x is, as before, the distance of the foot of the ladder from the wall, y is, instead of the height reached, the horizontal length of the wall, or the number of bricks in it, or the number of years since it was built, any change in x would naturally cause no change whatever in y; in this case $\dfrac{dy}{dx}$ has no meaning whatever, and it is not possible to find an expression for it. Whenever we use differentials dx, dy, dz, etc., the existence of some kind of relation between x, y, z, etc., is implied, and this relation is called a "function" in x, y, z, etc.; the two expressions given above, for instance, namely, $\dfrac{y}{x} = \tan 30°$ and $x^2 + y^2 = l^2$, are functions of x and y. Such expressions contain implicitly (that is, contain without distinctly showing it) the means of expressing either x in terms of y or y in terms of x, and for this reason they are called *implicit functions in x and y*; they can be respectively put into the forms

$$y = x \tan 30° \quad \text{or} \quad x = \frac{y}{\tan 30°}$$

and

$$y = \sqrt{l^2 - x^2} \quad \text{or} \quad x = \sqrt{l^2 - y^2}.$$

These last expressions state explicitly the value of x in terms of y, or of y in terms of x, and they are for this reason called *explicit functions* of x or y. For example $x^2 + 3 = 2y - 7$ is an implicit function of x and y; it may be written $y = \dfrac{x^2 + 10}{2}$ (explicit function of x) or $x = \sqrt{2y - 10}$ (explicit function of y). We see that an explicit function in x, y, z, etc., is simply something the value of which changes when x, y, z, etc., are changing, either one at the time or several together. Because of this, the value of the explicit function is called the *dependent variable,* as it depends on the value of the other variable quantities in the function; these other variables are called the *independent variables* because their value is not determined from the value assumed by the function. For example, if $u = x^2 \sin \theta$, x and θ are the independent variables, and u is the dependent variable.

Sometimes the exact relation between several quantities x, y, z

either is not known or it is not convenient to state it; it is only known, or convenient to state, that there is some sort of relation between these variables, so that one cannot alter either x or y or z singly without affecting the other quantities; the existence of a function in x, y, z is then indicated by the notation $F(x, y, z)$ (implicit function) or by $x = F(y, z)$, $y = F(x, z)$ or $z = F(x, y)$ (explicit function). Sometimes the letter f or ϕ is used instead of F, so that $y = F(x)$, $y = f(x)$ and $y = \phi(x)$ all mean the same thing, namely, that the value of y depends on the value of x in some way which is not stated.

We call the ratio $\dfrac{dy}{dx}$, "the *differential coefficient* of y with respect to x". It is a solemn scientific name for this very simple thing. But we are not going to be frightened by solemn names, when the things themselves are so easy. Instead of being frightened we will simply pronounce a brief curse on the stupidity of giving long crack-jaw names; and, having relieved out minds, will go on to the simple thing itself, namely the ratio $\dfrac{dy}{dx}$.[2]

In ordinary algebra which you learned at school, you were always hunting some unknown quantity which you called x or y; or sometimes there were two unknown quantities to be hunted for simultaneously. You have now to learn to go hunting in a new way; the fox being now neither x nor y. Instead of this you have to hunt for this curious cub called $\dfrac{dy}{dx}$. The process of finding the value of $\dfrac{dy}{dx}$ is called "differentiating". But, remember, what is wanted is the value of this ratio when both dy and dx are themselves infinitely small. The true value of the derivative is that to which it approximates in the limiting case when each of them is considered as infinitesimally minute.

Let us now learn how to go in quest of $\dfrac{dy}{dx}$.

2. I have let stand here Thompson's justified criticism of the term "differential coefficient," a term in use when he wrote his book. The term was later replaced by the simpler word "derivative." Henceforth in this book it will be called a derivative.—M.G.

NOTE TO CHAPTER III
How to Read Derivatives

It will never do to fall into the schoolboy error of thinking that dx means d times x, for d is not a factor—it means "an element of" or "a bit of" whatever follows. One reads dx thus: "dee-eks".

In case the reader has no one to guide him in such matters it may here be simply said that one reads derivatives in the following way. The derivative

$\dfrac{dy}{dx}$ is read *"dee-wy dee-eks"*, or *"dee-wy over dee-eks"*.

So also $\dfrac{du}{dt}$ is read *"dee-you dee-tee"*.

Second derivatives will be met with later on. They are like this: $\dfrac{d^2y}{dx^2}$; which is read *"dee-two-wy over dee-eks-squared"*, and it means that the operation of differentiating y with respect to x has been (or has to be) performed twice over.

Another way of indicating that a function has been differentiated is by putting an accent sign to the symbol of the function. Thus if $y = F(x)$, which means that y is some unspecified function of x, we may write $F'(x)$ instead of $\dfrac{d(F(x))}{dx}$. Similarly, $F''(x)$ will mean that the original function $F(x)$ has been differentiated twice over with respect to x.[3]

3. Newton called a variable a "fluent," and a derivative a "fluxion" because its value flowed or fluctuated continuously. In Chapter 10 Thompson tells how Newton indicated a first derivative by putting a dot over a term, a second derivative by putting two dots, a third derivative by three dots, and so on.

Morris Kline, in his two-volume *Calculus* (1967), is the only mathematician I know of in recent times who adopted Newton's dot notation for derivatives. Physicists, however, often use dot notation to denote differentiation with respect to time.—M.G.

Chapter IV

SIMPLEST CASES

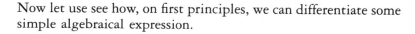

Now let use see how, on first principles, we can differentiate some simple algebraical expression.

Case 1.

Let us begin with the simple expression $y = x^2$.[1] Now remember that the fundamental notion about the calculus is the idea of *growing*. Mathematicians call it *varying*. Now as y and x^2 are equal to one another, it is clear that if x grows, x^2 will also grow. And if x^2 grows, then y will also grow. What we have got to find out is the proportion between the growing of y and the growing of x. In other words, our task is to find out the ratio between dy and dx, or, in brief, to find the value of $\dfrac{dy}{dx}$.

Let x, then, grow a little bit bigger and become $x + dx$; similarly, y will grow a bit bigger and will become $y + dy$. Then, clearly, it will still be true that the enlarged y will be equal to the square of the enlarged x. Writing this down, we have:

$$y + dy = (x + dx)^2$$

Doing the squaring we get:

$$y + dy = x^2 + 2x \cdot dx + (dx)^2$$

What does $(dx)^2$ mean? Remember that dx meant a bit—a little bit—of x. Then $(dx)^2$ will mean a little bit of a little bit of

1. The graph of this equation is a parabola.—M.G.

x^2; that is, as explained above, it is a small quantity of the second order of smallness. It may therefore be discarded as quite negligible in comparison with the other terms. Leaving it out, we then have:

$$y + dy = x^2 + 2x \cdot dx$$

Now $y = x^2$; so let us subtract this from the equation and we have left

$$dy = 2x \cdot dx$$

Dividing across by dx, we find

$$\frac{dy}{dx} = 2x$$

Now *this** is what we set out to find. The ratio of the growing of y to the growing of x is, in the case before us, found to be $2x$.

Numerical Example.
Suppose $x = 100$ and therefore $y = 10,000$. Then let x grow till it becomes 101 (that is, let $dx = 1$). Then the enlarged y will be $101 \times 101 = 10,201$. But if we agree that we may ignore small quantities of the second order, 1 may be rejected as compared with 10,000; so we may round off the enlarged y to 10,200; y has

---

*N.B.—This ratio $\frac{dy}{dx}$ is the result of differentiating y with respect to x. Differentiating means finding the derivative. Suppose we had some other function of x, as, for example, $u = 7x^2 + 3$. Then if we were told to differentiate this with respect to x, we should have to find $\frac{du}{dx}$, or, what is the same thing, $\frac{d(7x^2 + 3)}{dx}$. On the other hand, we may have a case in which time was the independent variable, such as this: $y = b + \frac{1}{2}at^2$. Then, if we were told to differentiate it, that means we must find its derivative with respect to t. So that then our business would be to try to find $\frac{dy}{dt}$, that is, to find $\frac{d(b + \frac{1}{2}at^2)}{dt}$.

grown from 10,000 to 10,200; the bit added on is dy, which is therefore 200.

$\dfrac{dy}{dx} = \dfrac{200}{1} = 200$. According to the algebra-working of the previous paragraph, we find $\dfrac{dy}{dx} = 2x$. And so it is; for $x = 100$ and

$2x = 200$.

But, you will say, we neglected a whole unit.

Well, try again, making dx a still smaller bit.

Try $dx = \frac{1}{10}$. Then $x + dx = 100.1$, and

$$(x + dx)^2 = 100.1 \times 100.1 = 10,020.01$$

Now the last figure 1 is only one-millionth part of the 10,000, and is utterly negligible; so we may take 10,020 without the little decimal at the end.[2] And this makes $dy = 20$; and $\dfrac{dy}{dx} = \dfrac{20}{0.1} = 200$, which is still the same as $2x$.

2. Many writers of calculus texts prefer to use the Greek letter delta, Δ, in place of d to stand for an increment that is small enough to be taken as zero. A derivative is defined as:

$$\frac{dy}{dx} = \lim_{\Delta x \to 0} \frac{f(x + \Delta x) - f(x)}{\Delta x}$$

This expresses the limit when Δx diminishes to zero. For example, if $f(x) = 2$, the formula becomes:

$$\frac{\Delta y}{\Delta x} = \frac{2 - 2}{\Delta x} = \frac{0}{\Delta x}$$

Since Δx goes to zero, the derivative of 2 is zero, and its graph is a horizontal line.

If $f(x) = 2x$, the formula gives $2\Delta x / \Delta x$. Because Δx goes to zero, the derivative of $2x$ is 2, and the function graphs as a straight line sloping upward.

Thompson does not use the Δ notation. Indeed, he avoids altogether the notion of limits. But no harm is done. It is easy to translate Thompson's technique of "exhausting" ever decreasing increments to the point where they can be "thrown away" into today's way of defining derivatives as limits.—M.G.

Case 2.

Try differentiating $y = x^3$ in the same way.

We let y grow to $y + dy$, while x grows to $x + dx$.

Then we have

$$y + dy = (x + dx)^3$$

Doing the cubing we obtain

$$y + dy = x^3 + 3x^2 \cdot dx + 3x(dx)^2 + (dx)^3$$

Now we know that we may neglect small quantities of the second and third orders; since, when dy and dx are both made infinitely small, $(dx)^2$ and $(dx)^3$ will become infinitely smaller by comparison. So, regarding them as negligible, we have left:

$$y + dy = x^3 + 3x^2 \cdot dx$$

But $y = x^3$; and, subtracting this, we have:

$$dy = 3x^2 \cdot dx$$

and

$$\frac{dy}{dx} = 3x^2$$

Case 3.

Try differentiating $y = x^4$. Starting as before by letting both y and x grow a bit, we have:

$$y + dy = (x + dx)^4$$

Working out the raising to the fourth power, we get

$$y + dy = x^4 + 4x^3dx + 6x^2(dx)^2 + 4x(dx)^3 + (dx)^4$$

Then, striking out the terms containing all the higher powers of dx, as being negligible by comparison, we have

$$y + dy = x^4 + 4x^3dx$$

Subtracting the original $y = x^4$, we have left

$$dy = 4x^3dx, \quad \text{and} \quad \frac{dy}{dx} = 4x^3$$

Now all these cases are quite easy. Let us collect the results to see if we can infer any general rule. Put them in two columns, the values of y in one and the corresponding values found for $\dfrac{dy}{dx}$ in the other: thus

y	$\dfrac{dy}{dx}$
x^2	$2x$
x^3	$3x^2$
x^4	$4x^3$

Just look at these results: the operation of differentiating appears to have had the effect of diminishing the power of x by 1 (for example in the last case reducing x^4 to x^3), and at the same time multiplying by a number (the same number in fact which originally appeared as the power). Now, when you have once seen this, you might easily conjecture how the others will run. You would expect that differentiating x^5 would give $5x^4$, or differentiating x^6 would give $6x^5$. If you hesitate, try one of these, and see whether the conjecture comes right.

Try $y = x^5$.

Then $y + dy = (x + dx)^5 = x^5 + 5x^4dx + 10x^3(dx)^2 + 10x^2(dx)^3$
$$+ 5x(dx)^4 + (dx)^5.$$

Neglecting all the terms containing small quantities of the higher orders, we have left

$$y + dy = x^5 + 5x^4dx$$

and subtracting $y = x^5$ leaves us

$$dy = 5x^4dx$$

whence $\qquad \dfrac{dy}{dx} = 5x^4$, exactly as we supposed.

Following out logically our observation, we should conclude that if we want to deal with any higher power—call it x^n—we could tackle it in the same way.

Let $$y = x^n$$

then we should expect to find that

$$\frac{dy}{dx} = nx^{n-1}$$

For example, let $n = 8$, then $y = x^8$; and differentiating it would give $\frac{dy}{dx} = 8x^7$.

And, indeed, the rule that differentiating x^n gives as the result nx^{n-1} is true for all cases where n is a whole number and positive. [Expanding $(x + dx)^n$ by the binomial theorem will at once show this.] But the question whether it is true for cases where n has negative or fractional values requires further consideration.

Case of a Negative Exponent.
Let $y = x^{-2}$. Then proceed as before:

$$y + dy = (x + dx)^{-2}$$
$$= x^{-2}\left(1 + \frac{dx}{x}\right)^{-2}$$

Expanding this by the binomial theorem, we get

$$= x^{-2}\left[1 - \frac{2dx}{x} + \frac{2(2+1)}{1 \times 2}\left(\frac{dx}{x}\right)^2 - \ldots\right]$$
$$= x^{-2} - 2x^{-3} \cdot dx + 3x^{-4}(dx)^2 - 4x^{-5}(dx)^3 + \text{etc.}$$

So, neglecting the small quantities of higher orders of smallness, we have:

$$y + dy = x^{-2} - 2x^{-3} \cdot dx$$

Subtracting the original $y = x^{-2}$, we find

$$dy = -2x^{-3}dx$$

$$\frac{dy}{dx} = -2x^{-3}$$

And this is still in accordance with the rule inferred above.

Case of a Fractional Exponent.
Let $y = x^{\frac{1}{2}}$. Then, as before,

$$y + dy = (x + dx)^{\frac{1}{2}} = x^{\frac{1}{2}}\left(1 + \frac{dx}{x}\right)^{\frac{1}{2}} = \sqrt{x}\left(1 + \frac{dx}{x}\right)^{\frac{1}{2}}$$

$$= \sqrt{x} + \frac{1}{2}\frac{dx}{\sqrt{x}} - \frac{1}{8}\frac{(dx)^2}{x\sqrt{x}} + \text{terms with higher powers of } dx.$$

Subtracting the original $y = x^{\frac{1}{2}}$, and neglecting higher powers we have left:

$$dy = \frac{1}{2}\frac{dx}{\sqrt{x}} = \frac{1}{2}x^{-\frac{1}{2}} \cdot dx$$

and $\quad \dfrac{dy}{dx} = \dfrac{1}{2}x^{-\frac{1}{2}}$. This agrees with the general rule.

Summary. Let us see how far we have got. We have arrived at the following rule: To differentiate x^n, multiply it by the exponent and reduce the exponent by one, so giving us nx^{n-1} as the result.[3]

---

3. This rule is today known as the "power rule." It is the rule most frequently used in differentiating low-order functions.—M.G.

EXERCISES I

Differentiate the following:

(1) $y = x^{13}$ (2) $y = x^{-\frac{3}{2}}$

(3) $y = x^{2a}$ (4) $u = t^{2.4}$

(5) $z = \sqrt[3]{u}$ (6) $y = \sqrt[3]{x^{-5}}$

(7) $u = \sqrt[5]{\dfrac{1}{x^8}}$ (8) $y = 2x^a$

(9) $y = \sqrt[q]{x^3}$ (10) $y = \sqrt[n]{\dfrac{1}{x^m}}$

You have now learned how to differentiate powers of x. How easy it is!

Chapter V

NEXT STAGE. WHAT TO DO WITH CONSTANTS

———— ∞≪≫∞ ————

In our equations we have regarded x as growing, and as a result of x being made to grow y also changed its value and grew. We usually think of x as a quantity that we can vary; and, regarding the variation of x as a sort of *cause*, we consider the resulting variation of y as an *effect*. In other words, we regard the value of y as depending on that of x. Both x and y are variables, but x is the one that we operate upon, and y is the "dependent variable". In all the preceding chapters we have been trying to find out rules for the proportion which the dependent variation in y bears to the variation independently made in x.

Our next step is to find out what effect on the process of differentiating is caused by the presence of *constants*, that is, of numbers which don't change when x or y changes its value.

Added Constants.

Let us begin with a simple case of an added constant, thus:

Let $$y = x^3 + 5$$

Just as before, let us suppose x to grow to $x + dx$ and y to grow to $y + dy$.

Then:
$$y + dy = (x + dx)^3 + 5$$
$$= x^3 + 3x^2dx + 3x(dx)^2 + (dx)^3 + 5$$

Neglecting the small quantities of higher orders, this becomes

$$y + dy = x^3 + 3x^2 \cdot dx + 5$$

Subtract the original $y = x^3 + 5$, and we have left:

$$dy = 3x^2 dx$$

$$\frac{dy}{dx} = 3x^2$$

So the 5 has quite disappeared. It added nothing to the growth of x, and does not enter into the derivative. If we had put 7, or 700, or any other number, instead of 5, it would have disappeared. So if we take the letter a, or b, or c to represent any constant, it will simply disappear when we differentiate.

If the additional constant had been of negative value, such as -5 or $-b$, it would equally have disappeared.

Multiplied Constants.
Take as a simple experiment this case:
   Let $y = 7x^2$
Then on proceeding as before we get:

$$y + dy = 7(x + dx)^2$$
$$= 7\{x^2 + 2x \cdot dx + (dx)^2\}$$
$$= 7x^2 + 14x \cdot dx + 7(dx)^2$$

Then, subtracting the original $y = 7x^2$, and neglecting the last term, we have

$$dy = 14x \cdot dx$$

$$\frac{dy}{dx} = 14x$$

Let us illustrate this example by working out the graphs of the equations $y = 7x^2$ and $\frac{dy}{dx} = 14x$, by assigning to x a set of successive values, 0, 1, 2, 3, etc., and finding the corresponding values of y and of $\frac{dy}{dx}$.

These values we tabulate as follows:

x	0	1	2	3	4	5	−1	−2	−3
y	0	7	28	63	112	175	7	28	63
$\dfrac{dy}{dx}$	0	14	28	42	56	70	−14	−28	−42

Now plot these values to some convenient scale, and we obtain the two curves, Figs. 6 and 6*a*.

Carefully compare the two figures, and verify by inspection that the height of the ordinate of the derived curve, Fig. 6*a*, is porportional to the *slope* of the original curve, Fig. 6, at the corresponding value of *x*. To the left of the origin, where the original curve slopes negatively (that is, downward from left to right) the corresponding ordinates of the derived curve are negative.

Now, if we look back at previous pages, we shall see that simply differentiating x^2 gives us $2x$. So that the derivative of $7x^2$ is just 7 times as big as that of x^2. If we had taken $8x^2$, the derivative would have come out eight times as great as that of x^2. If we put $y = ax^2$, we shall get

$$\frac{dy}{dx} = a \times 2x$$

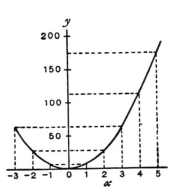

FIG. 6. Graph of $y = 7x^2$.

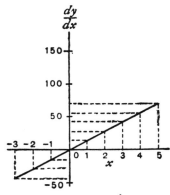

FIG. 6a. Graph of $\dfrac{dy}{dx} = 14x$.

If we had begun with $y = ax^n$, we should have had

$$\frac{dy}{dx} = a \times nx^{n-1}$$

So that any mere multiplication by a constant reappears as a mere multiplication when the thing is differentiated. And what is true about multiplication is equally true about *division:* for if, in the example above, we had taken as the constant $\frac{1}{7}$ instead of 7, we should have had the same $\frac{1}{7}$ come out in the result after differentiation.

Some Further Examples.

The following further examples, fully worked out, will enable you to master completely the process of differentiation as applied to ordinary algebraical expressions, and enable you to work out by yourself the examples given at the end of this chapter.

(1) Differentiate $y = \dfrac{x^5}{7} - \dfrac{3}{5}$

$-\dfrac{3}{5}$ is an added constant and vanishes.

We may then write at once

$$\frac{dy}{dx} = \frac{1}{7} \times 5 \times x^{5-1}$$

or
$$\frac{dy}{dx} = \frac{5}{7} x^4$$

(2) Differentiate $y = a\sqrt{x} - \dfrac{1}{2}\sqrt{a}$

The term $-\dfrac{1}{2}\sqrt{a}$ vanishes, being an added constant; and as $a\sqrt{x}$, in the index form, is written $ax^{\frac{1}{2}}$, we have

$$\frac{dy}{dx} = a \times \frac{1}{2} \times x^{\frac{1}{2}-1} = \frac{a}{2} \times x^{-\frac{1}{2}}$$

or
$$\frac{dy}{dx} = \frac{a}{2\sqrt{x}}$$

(3) The volume of a right circular cylinder of radius r and height h is given by the formula $V = \pi r^2 h$. Find the rate of variation of volume with the radius when $r = 5.5$ in. and $h = 20$ in. If $r = h$, find the dimensions of the cylinder so that a change of 1 in. in radius causes a change of 400 cubic inches in the volume.

The rate of variation of V with regard to r is

$$\frac{dV}{dr} = 2\pi r h$$

If $r = 5.5$ in. and $h = 20$ in. this becomes 691.2. It means that a change of radius of 1 inch will cause a change of volume of 691.2 cubic inches. This can be easily verified, for the volumes with $r = 5$ and $r = 6$ are 1570.8 cubic inches and 2262 cubic inches respectively, and $2262 - 1570.8 = 691.2$.

Also, if $h = r$, and h remains constant,

$$\frac{dV}{dr} = 2\pi r^2 = 400 \quad \text{and} \quad r = h = \sqrt{\frac{400}{2\pi}} = 7.98 \text{ in.}$$

If, however, $h = r$ and varies with r, then

$$\frac{dV}{dr} = 3\pi r^2 = 400 \quad \text{and} \quad r = h = \sqrt{\frac{400}{3\pi}} = 6.51 \text{ in.}$$

(4) The reading θ of a Féry's Radiation pyrometer is related to the centigrade temperature t of the observed body by the relation $\theta/\theta_1 = (t/t_1)^4$, where θ_1 is the reading corresponding to a known temperature t_1 of the observed body.

Compare the sensitivity of the pyrometer at temperatures 800° C., 1000° C., 1200° C., given that it read 25 when the temperature was 1000° C.

The sensitivity is the rate of variation of the reading with the temperature, that is, $\dfrac{d\theta}{dt}$. The formula may be written

$$\theta = \frac{\theta_1}{t_1{}^4} t^4 = \frac{25 t^4}{1000^4}$$

and we have $\qquad \dfrac{d\theta}{dt} = \dfrac{100 t^3}{1000^4} = \dfrac{t^3}{10,000,000,000}$

When $t = 800$, 1000 and 1200, we get $\dfrac{d\theta}{dt} = 0.0512$, 0.1 and 0.1728 respectively.

The sensitivity is approximately doubled from 800° to 1000°, and becomes three-quarters as great again up to 1200°.

EXERCISES II

Differentiate the following:

(1) $y = ax^3 + 6$

(2) $y = 13x^{\frac{3}{2}} - c$

(3) $y = 12x^{\frac{1}{2}} + c^{\frac{1}{2}}$

(4) $y = c^{\frac{1}{2}}x^{\frac{1}{2}}$

(5) $u = \dfrac{az^n - 1}{c}$

(6) $y = 1.18t^2 + 22.4$

Make up some other examples for yourself, and try your hand at differentiating them.

(7) If l_t and l_0 be the lengths of a rod of iron at the temperatures $t°$ C. and 0° C. respectively, then $l_t = l_0 (1 + 0.000012t)$. Find the change of length of the rod per degree centigrade.

(8) It has been found that if c be the candle power of an incandescent electric lamp, and V be the voltage, $c = aV^b$, where a and b are constants.

Find the rate of change of the candle power with the voltage, and calculate the change of candle power per volt at 80, 100 and 120 volts in the case of a lamp for which $a = 0.5 \times 10^{-10}$ and $b = 6$.

(9) The frequency n of vibration of a string of diameter D, length L and specific gravity σ, stretched with a force T, is given by

$$n = \frac{1}{DL} \sqrt{\frac{gT}{\pi\sigma}}$$

Find the rate of change of the frequency when D, L, σ and T are varied singly.

(10) The greatest external pressure P which a tube can support without collapsing is given by

$$P = \left(\frac{2E}{1 - \sigma^2} \right) \frac{t^3}{D^3}$$

where E and σ are constants, t is the thickness of the tube and D is its diameter. (This formula assumes that $4t$ is small compared to D.)

Compare the rate at which P varies for a small change of thickness and for a small change of diameter taking place separately.

(11) Find, from first principles, the rate at which the following vary with respect to a change in radius:

(a) the circumference of a circle of radius r;
(b) the area of a circle of radius r;
(c) the lateral area of a cone of slant dimension l;
(d) the volume of a cone of radius r and height h;
(e) the area of a sphere of radius r;
(f) the volume of a sphere of radius r.

Chapter VI

SUMS, DIFFERENCES, PRODUCTS, AND QUOTIENTS[1]

We have learned how to differentiate simple algebraical functions such as $x^2 + c$ or ax^4, and we have now to consider how to tackle the *sum* of two or more functions.

For instance, let

$$y = (x^2 + c) + (ax^4 + b)$$

what will its $\dfrac{dy}{dx}$ be? How are we to go to work on this new job?

The answer to this question is quite simple: just differentiate them, one after the other, thus:

$$\frac{dy}{dx} = 2x + 4ax^3$$

If you have any doubt whether this is right, try a more general case, working it by first principles. And this is the way.

Let $y = u + v$, where u is any function of x, and v any other function of x. Then, letting x increase to $x + dx$, y will increase to $y + dy$; and u will increase to $u + du$; and v to $v + dv$.

And we shall have:

$$y + dy = u + du + v + dv$$

---

1. The rules given in this chapter are today known as the sum rule, the difference rule, the product rule, and the quotient rule.—M.G.

Subtracting the original $y = u + v$, we get

$$dy = du + dv$$

and dividing through by dx, we get:

$$\frac{dy}{dx} = \frac{du}{dx} + \frac{dv}{dx}$$

This justifies the procedure. You differentiate each function separately and add the results. So if now we take the example of the preceding paragraph, and put in the values of the two functions, we shall have, using the notation shown.

$$\frac{dy}{dx} = \frac{d(x^2 + c)}{dx} + \frac{d(ax^4 + b)}{dx}$$

$$= 2x \qquad\quad + 4ax^3$$

exactly as before.

If there were three functions of x, which we may call u, v and w, so that

$$y = u + v + w$$

then

$$\frac{dy}{dx} = \frac{du}{dx} + \frac{dv}{dx} + \frac{dw}{dx}$$

As for the rule about *subtraction,* it follows at once; for if the function v had itself had a negative sign, its derivative would also be negative; so that by differentiating

$$y = u - v$$

we should get

$$\frac{dy}{dx} = \frac{du}{dx} - \frac{dv}{dx}$$

But when we come to do with *Products,* the thing is not quite so simple.

Suppose we were asked to differentiate the expression

$$y = (x^2 + c) \times (ax^4 + b)$$

what are we to do? The result will certainly *not* be $2x \times 4ax^3$; for it is easy to see that neither $c \times ax^4$, nor $x^2 \times b$, would have been taken into that product.

Now there are two ways in which we may go to work.

First Way.   Do the multiplying first, and, having worked it out, then differentiate.

Accordingly, we multiply together $x^2 + c$ and $ax^4 + b$.

This gives $ax^6 + acx^4 + bx^2 + bc$.

Now differentiate, and we get:

$$\frac{dy}{dx} = 6ax^5 + 4acx^3 + 2bx$$

Second Way.   Go back to first principles, and consider the equation

$$y = u \times v$$

where u is one function of x, and v is any other function of x. Then, if x grows to be $x + dx$; and y to $y + dy$; and u becomes $u + du$; and v becomes $v + dv$, we shall have:

$$y + dy = (u + du) \times (v + dv)$$

$$= u \cdot v + u \cdot dv + v \cdot du + du \cdot dv$$

Now $du \cdot dv$ is a small quantity of the second order of smallness, and therefore in the limit may be discarded, leaving

$$y + dy = u \cdot v + u \cdot dv + v \cdot du$$

Then, subtracting the original $y = u \cdot v$, we have left

$$dy = u \cdot dv + v \cdot du$$

and, dividing through by dx, we get the result:

$$\frac{dy}{dx} = u\frac{dv}{dx} + v\frac{du}{dx}$$

This shows that our instructions will be as follows: *To differentiate the product of two functions, multiply each function by the derivative of the other, and add together the two products so obtained.*

You should note that this process amounts to the following:

Treat u as constant while you differentiate *v*; then treat *v* as constant while you differentiate *u*; and the whole derivative $\frac{dy}{dx}$ will be the sum of the results of these two treatments.

Now, having found this rule, apply it to the concrete example which was considered above.

We want to differentiate the product

$$(x^2 + c) \times (ax^4 + b)$$

Call $\qquad (x^2 + c) = u;$ and $(ax^4 + b) = v$

Then, by the general rule just established, we may write:

$$\frac{dy}{dx} = (x^2 + c)\frac{d(ax^4 + b)}{dx} + (ax^4 + b)\frac{d(x^2 + c)}{dx}$$

$$= (x^2 + c)4ax^3 \qquad + (ax^4 + b)2x$$

$$= 4ax^5 + 4acx^3 \qquad + 2ax^5 + 2bx$$

$$\frac{dy}{dx} = 6ax^5 + 4acx^3 \qquad + 2bx$$

Exactly as before.

Lastly, we have to differentiate *quotients*.

Think of this example, $y = \dfrac{bx^5 + c}{x^2 + a}$. In such a case it is no use to try to work out the division beforehand, because $x^2 + a$ will not divide into $bx^5 + c$, neither have they any common factor. So there is nothing for it but to go back to first principles, and find a rule.

So we will put $\qquad\qquad y = \dfrac{u}{v}$

where *u* and *v* are two different functions of the independent variable *x*. Then, when *x* becomes $x + dx$, *y* will become $y + dy$; and *u* will become $u + du$; and *v* will become $v + dv$. So then

$$y + dy = \frac{u + du}{v + dv}$$

Now perform the algebraic division, thus:

$$v + dv \;\Big|\; u + du \;\Big|\; \dfrac{u}{v} + \dfrac{du}{v} - \dfrac{u \cdot dv}{v^2}$$

$$\begin{array}{l}
u + \dfrac{u \cdot dv}{v} \\[2mm]
\hline
du - \dfrac{u \cdot dv}{v} \\[2mm]
du + \dfrac{du \cdot dv}{v} \\[2mm]
\hline
-\dfrac{u \cdot dv}{v} - \dfrac{du \cdot dv}{v} \\[2mm]
-\dfrac{u \cdot dv}{v} - \dfrac{u \cdot dv \cdot dv}{v^2} \\[2mm]
\hline
\qquad -\dfrac{du \cdot dv}{v} + \dfrac{u \cdot dv \cdot dv}{v^2}
\end{array}$$

As both these remainders are small quantities of the second order, they may be neglected, and the division may stop here, since any further remainders would be of still smaller magnitudes.

So we have got:

$$y + dy = \frac{u}{v} + \frac{du}{v} - \frac{u \cdot dv}{v^2}$$

which may be written

$$= \frac{u}{v} + \frac{v \cdot du - u \cdot dv}{v^2}$$

Now subtract the original $y = \dfrac{u}{v}$, and we have left:

$$dy = \frac{v \cdot du - u \cdot dv}{v^2}$$

whence

$$\frac{dy}{dx} = \frac{v\dfrac{du}{dx} - u\dfrac{dv}{dx}}{v^2}$$

This gives us our instructions as to *how to differentiate a quotient of two functions. Multiply the divisor function by the derivative of the dividend function; then multiply the dividend function by the derivative of the divisor function; and subtract the latter product from the former. Lastly, divide the difference by the square of the divisor function.*

Going back to our example $y = \dfrac{bx^5 + c}{x^2 + a}$

write $\qquad\qquad bx^5 + c = u; \quad \text{and} \quad x^2 + a = v$

Then $\quad \dfrac{dy}{dx} = \dfrac{(x^2 + a)\dfrac{d(bx^5 + c)}{dx} - (bx^5 + c)\dfrac{d(x^2 + a)}{dx}}{(x^2 + a)^2}$

$$= \frac{(x^2 + a)(5bx^4) - (bx^5 + c)(2x)}{(x^2 + a)^2}$$

$$\frac{dy}{dx} = \frac{3bx^6 + 5abx^4 - 2cx}{(x^2 + a)^2}$$

The working out of quotients is often tedious, but there is nothing difficult about it.

Some further examples fully worked out are given hereafter.

(1) Differentiate $\quad y = \dfrac{a}{b^2}x^3 - \dfrac{a^2}{b}x + \dfrac{a^2}{b^2}$

Being a constant, $\dfrac{a^2}{b^2}$ vanishes, and we have

$$\frac{dy}{dx} = \frac{a}{b^2} \times 3 \times x^{3-1} - \frac{a^2}{b} \times 1 \times x^{1-1}$$

But $x^{1-1} = x^0 = 1$; so we get:

$$\frac{dy}{dx} = \frac{3a}{b^2}x^2 - \frac{a^2}{b}$$

(2) Differentiate $\quad y = 2a\sqrt{bx^3} - \dfrac{3b\sqrt[3]{a}}{x} - 2\sqrt{ab}$

Putting x in the exponent form, we get

$$y = 2a\sqrt{b}\, x^{\frac{3}{2}} - 3b\sqrt[3]{a}\, x^{-1} - 2\sqrt{ab}$$

Now $\quad \dfrac{dy}{dx} = 2a\sqrt{b} \times \tfrac{3}{2} \times x^{\frac{3}{2}-1} - 3b\sqrt[3]{a} \times (-1) \times x^{-1-1}$

or, $\quad \dfrac{dy}{dx} = 3a\sqrt{bx} + \dfrac{3b\sqrt[3]{a}}{x^{2}}$

(3) Differentiate $\quad z = 1.8 \sqrt[3]{\dfrac{1}{\theta^{2}}} - \dfrac{4.4}{\sqrt[5]{\theta}} - 27.$

This may be written: $z = 1.8\theta^{-\frac{2}{3}} - 4.4\theta^{-\frac{1}{5}} - 27.$
The 27 vanishes, and we have

$$\dfrac{dz}{d\theta} = 1.8 \times \left(-\tfrac{2}{3}\right)\theta^{-\frac{2}{3}-1} - 4.4 \times \left(-\tfrac{1}{5}\right)\theta^{-\frac{1}{5}-1}$$

or, $\quad \dfrac{dz}{d\theta} = -1.2\theta^{-\frac{5}{3}} + 0.88\theta^{-\frac{6}{5}}$

or, $\quad \dfrac{dz}{d\theta} = \dfrac{0.88}{\sqrt[5]{\theta^{6}}} - \dfrac{1.2}{\sqrt[3]{\theta^{5}}}$

(4) Differentiate $\quad v = (3t^{2} - 1.2t + 1)^{3}.$

A direct way of doing this will be explained later; but we can nevertheless manage it now without any difficulty.

Developing the cube, we get

$$v = 27t^{6} - 32.4t^{5} + 39.96t^{4} - 23.328t^{3} + 13.32t^{2} - 3.6t + 1$$

hence

$$\dfrac{dv}{dt} = 162t^{5} - 162t^{4} + 159.84t^{3} - 69.984t^{2} + 26.64t - 3.6.$$

(5) Differentiate $\quad y = (2x - 3)(x + 1)^{2}.$

$$\frac{dy}{dx} = (2x - 3)\frac{d[(x + 1)(x + 1)]}{dx} + (x + 1)^2\frac{d(2x - 3)}{dx}$$

$$= (2x - 3)\left[(x + 1)\frac{d(x + 1)}{dx} + (x + 1)\frac{d(x + 1)}{dx}\right]$$

$$+ (x + 1)^2\frac{d(2x - 3)}{dx}$$

$$= 2(x + 1)[(2x - 3) + (x + 1)] = 2(x + 1)(3x - 2)$$

or, more simply, multiply out and then differentiate.

(6) Differentiate $y = 0.5x^3(x - 3)$.

$$\frac{dy}{dx} = 0.5\left[x^3\frac{d(x - 3)}{dx} + (x - 3)\frac{d(x^3)}{dx}\right]$$

$$= 0.5[x^3 + (x - 3) \times 3x^2] = 2x^3 - 4.5x^2$$

Same remarks as for preceding example.

(7) Differentiate $w = \left(\theta + \dfrac{1}{\theta}\right)\left(\sqrt{\theta} + \dfrac{1}{\sqrt{\theta}}\right)$

This may be written

$$w = (\theta + \theta^{-1})\left(\theta^{\frac{1}{2}} + \theta^{-\frac{1}{2}}\right)$$

$$\frac{dw}{d\theta} = (\theta + \theta^{-1})\frac{d\left(\theta^{\frac{1}{2}} + \theta^{-\frac{1}{2}}\right)}{d\theta} + \left(\theta^{\frac{1}{2}} + \theta^{-\frac{1}{2}}\right)\frac{d(\theta + \theta^{-1})}{d\theta}$$

$$= (\theta + \theta^{-1})\left(\tfrac{1}{2}\theta^{-\frac{1}{2}} - \tfrac{1}{2}\theta^{-\frac{3}{2}}\right) + \left(\theta^{\frac{1}{2}} + \theta^{-\frac{1}{2}}\right)(1 - \theta^{-2})$$

$$= \tfrac{1}{2}\left(\theta^{\frac{1}{2}} + \theta^{-\frac{3}{2}} - \theta^{-\frac{1}{2}} - \theta^{-\frac{5}{2}}\right) + \left(\theta^{\frac{1}{2}} + \theta^{-\frac{1}{2}} - \theta^{-\frac{3}{2}} - \theta^{-\frac{5}{2}}\right)$$

$$= \frac{3}{2}\left(\sqrt{\theta} - \frac{1}{\sqrt{\theta^5}}\right) + \frac{1}{2}\left(\frac{1}{\sqrt{\theta}} - \frac{1}{\sqrt{\theta^3}}\right)$$

This, again, could be obtained more simply by multiplying the two factors first, and differentiating afterwards. This is not, however, always possible; see, for instance, example 8 in Chapter XV, in which the rule for differentiating a product *must* be used.

(8) Differentiate $\quad y = \dfrac{a}{1 + a\sqrt{x} + a^2 x}$

$$\frac{dy}{dx} = \frac{\left(1 + ax^{\frac{1}{2}} + a^2 x\right) \times 0 - a\dfrac{d\left(1 + ax^{\frac{1}{2}} + a^2 x\right)}{dx}}{\left(1 + a\sqrt{x} + a^2 x\right)^2}$$

$$= -\frac{a\left(\frac{1}{2}ax^{-\frac{1}{2}} + a^2\right)}{(1 + ax^{\frac{1}{2}} + a^2 x)^2}$$

(9) Differentiate $\quad y = \dfrac{x^2}{x^2 + 1}.$

$$\frac{dy}{dx} = \frac{(x^2 + 1)2x - x^2 \times 2x}{(x^2 + 1)^2} = \frac{2x}{(x^2 + 1)^2}$$

(10) Differentiate $\quad y = \dfrac{a + \sqrt{x}}{a - \sqrt{x}}$

In the exponent form, $\quad y = \dfrac{a + x^{\frac{1}{2}}}{a - x^{\frac{1}{2}}}$

$$\frac{dy}{dx} = \frac{\left(a - x^{\frac{1}{2}}\right)\left(\frac{1}{2}x^{-\frac{1}{2}}\right) - \left(a + x^{\frac{1}{2}}\right)\left(-\frac{1}{2}x^{-\frac{1}{2}}\right)}{\left(a - x^{\frac{1}{2}}\right)^2} = \frac{a - x^{\frac{1}{2}} + a + x^{\frac{1}{2}}}{2\left(a - x^{\frac{1}{2}}\right)^2 x^{\frac{1}{2}}}$$

hence $\qquad \dfrac{dy}{dx} = \dfrac{a}{\left(a - \sqrt{x}\right)^2 \sqrt{x}}$

(11) Differentiate $\quad \theta = \dfrac{1 - a\sqrt[3]{t^2}}{1 + a\sqrt[2]{t^3}}$

Now $\qquad \theta = \dfrac{1 - at^{\frac{2}{3}}}{1 + at^{\frac{3}{2}}}$

$$\frac{d\theta}{dt} = \frac{\left(1 + at^{\frac{3}{2}}\right)\left(-\frac{2}{3}at^{-\frac{1}{3}}\right) - \left(1 - at^{\frac{2}{3}}\right) \times \frac{3}{2}at^{\frac{1}{2}}}{\left(1 + at^{\frac{3}{2}}\right)^2}$$

$$= \frac{5a^2\sqrt[6]{t^7} - \dfrac{4a}{\sqrt[3]{t}} - 9a\sqrt[2]{t}}{6\left(1 + a\sqrt[2]{t^3}\right)^2}$$

(12) A reservoir of square cross-section has sides sloping at an angle of 45° with the vertical. The side of the bottom is p feet in length, and water flows in the reservoir at the rate of c cubic feet per minute. Find an expression for the rate at which the surface of the water is rising at the instant its depth is h feet. Calculate this rate when $p = 17$, $h = 4$ and $c = 35$.

The volume of a frustum of pyramid of height H, and of bases A and a, is $V = \dfrac{H}{3}\left(A + a + \sqrt{Aa}\right)$. It is easily seen that, the slope being 45°, for a depth of h, the length of the side of the upper square surface of water is $(p + 2h)$ feet; thus $A = p^2$, $a = (p + 2h)^2$ and the volume of the water is

$$\tfrac{1}{3}h\{p^2 + p(p + 2h) + (p + 2h)^2\} \text{ cubic feet}$$
$$= p^2h + 2ph^2 + \tfrac{4}{3}h^3 \text{ cubic feet}$$

Now, if t be the time in minutes taken for this volume of water to flow in,

$$ct = p^2h + 2ph^2 + \tfrac{4}{3}h^3$$

From this relation we have the rate at which h increases with t, that is $\dfrac{dh}{dt}$, but as the above expression is in the form, $t = $ function of h, rather than $h = $ function of t, it will be easier to find $\dfrac{dt}{dh}$ and then invert the result, for

$$\frac{dt}{dh} \times \frac{dh}{dt} = 1$$

Hence, since c and p are constants, and

$$ct = p^2 h + 2ph^2 + \tfrac{4}{3}h^3$$

$$c\frac{dt}{dh} = p^2 + 4ph + 4h^2 = (p + 2h)^2$$

so that $\dfrac{dh}{dt} = \dfrac{c}{(p+2h)^2}$, which is the required expression.

When $p = 17$, $h = 4$ and $c = 35$; this becomes 0.056 feet per minute.

(13) The absolute pressure, in atmospheres, P, of saturated steam at the temperature $t°$ C. is $P = \left(\dfrac{40+t}{140}\right)^5$ as long as t is above 80°. Find the rate of variation of the pressure with the temperature at 100° C.

Since $\qquad P = \left(\dfrac{40+t}{140}\right)^5$; $\quad \dfrac{dP}{dt} = \dfrac{5(40+t)^4}{(140)^5}$

so that when $t = 100$,

$$\frac{dP}{dt} = \frac{5 \times (140)^4}{(140)^5} = \frac{5}{140} = \frac{1}{28} = 0.036$$

Thus, the rate of variation of the pressure is, when $t = 100$, 0.036 atmosphere per degree centigrade change of temperature.

EXERCISES III

(1) Differentiate

(a) $u = 1 + x + \dfrac{x^2}{1 \times 2} + \dfrac{x^3}{1 \times 2 \times 3} + \dots$

(b) $y = ax^2 + bx + c$ $\qquad$ (c) $y = (x+a)^2$

(d) $y = (x+a)^3$

(2) If $w = at - \frac{1}{2}bt^2$, find $\dfrac{dw}{dt}$

(3) Find the derivative of

$$y = \left(x + \sqrt{-1}\right) \times \left(x - \sqrt{-1}\right)$$

(4) Differentiate

$$y = (197x - 34x^2) \times (7 + 22x - 83x^3)$$

(5) If $x = (y + 3) \times (y + 5)$, find $\dfrac{dx}{dy}$.

(6) Differentiate $y = 1.3709x \times (112.6 + 45.202x^2)$.

Find the derivatives of

(7) $y = \dfrac{2x + 3}{3x + 2}$

(8) $y = \dfrac{1 + x + 2x^2 + 3x^3}{1 + x + 2x^2}$

(9) $y = \dfrac{ax + b}{cx + d}$

(10) $y = \dfrac{x^n + a}{x^{-n} + b}$

(11) The temperature t of the filament of an incandescent electric lamp is connected to the current passing through the lamp by the relation

$$C = a + bt + ct^2$$

Find an expression giving the variation of the current corresponding to a variation of temperature.

(12) The following formulae have been proposed to express the relation between the electric resistance R of a wire at the temperature $t°$ C., and the resistance R_0 of that same wire at $0°$ centigrade, a and b being constants.

$$R = R_0(1 + at + bt^2)$$
$$R = R_0\left(1 + at + b\sqrt{t}\right)$$
$$R = R_0(1 + at + bt^2)^{-1}$$

Find the rate of variation of the resistance with regard to temperature as given by each of these formulae.

(13) The electromotive force E of a certain type of standard cell has been found to vary with the temperature t according to the relation

$$E = 1.4340[1 - 0.000814(t - 15) \\ + 0.000007(t - 15)^2] \text{ volts}$$

Find the change of electromotive force per degree, at 15°, 20° and 25°.

(14) The electromotive force necessary to maintain an electric arc of length l with a current intensity i has been found to be

$$E = a + bl + \frac{c + kl}{i}$$

where a, b, c, k are constants.

Find an expression for the variation of the electromotive force (a) with regard to the length of the arc; (b) with regard to the strength of the current.

Chapter VII

SUCCESSIVE DIFFERENTIATION

———∞∞———

Let us try the effect of repeating several times over the operation of differentiating a function. Begin with a concrete case.

Let $y = x^5$.

First differentiation, $5x^4$.
Second differentiation, $5 \times 4x^3$ $= 20x^3$.
Third differentiation, $5 \times 4 \times 3x^2$ $= 60x^2$.
Fourth differentiation, $5 \times 4 \times 3 \times 2x$ $= 120x$.
Fifth differentiation, $5 \times 4 \times 3 \times 2 \times 1 = 120$.
Sixth differentiation, $= 0.^1$

1. When applied to an object moving at a constant rate, the first derivative gives its change of position per second. If the object is accelerating, the second derivative gives the rate at which the first derivative is changing, that is, its change of position per second per second. If the acceleration is changing, the third derivative gives the rate at which the second derivative is changing, namely the change of position per second per second per second. Physicists call such a change a "jerk," such as the way an old car jerks if there is too sudden a change in how it is accelerating.

Second derivatives, in which time is the independent variable, turn up everywhere in physics, less often in other sciences. In economics, a second derivative can express the rate at which a worker's annual increase in wages is increasing (or decreasing). Third derivatives are also useful in many branches of physics. Beyond the third, higher order derivatives are seldom needed. This testifies to the fortunate fact that the universe seems to favor simplicity in its fundamental laws.—M.G.

There is a certain notation, with which we are already acquainted, used by some writers, that is very convenient. This is to employ the general symbol $f(x)$ for any function of x. Here the symbol $f(\ )$ is read as "function of", without saying what particular function is meant. So the statement $y = f(x)$ merely tells us that y is a function of x, it may be x^2 or ax^n, or $\cos x$ or any other complicated function of x.

The corresponding symbol for the derivative is $f'(x)$, which is simpler to write than $\dfrac{dy}{dx}$. This is called the "derived function" of x.

Suppose we differentiate over again, we shall get the "second derived function" or second derivative which is denoted by $f''(x)$; and so on.

Now let us generalize.

Let $y = f(x) = x^n$.

First differentiation, $f'(x) = nx^{n-1}$.
Second differentiation, $f''(x) = n(n-1)x^{n-2}$.
Third differentiation, $f'''(x) = n(n-1)(n-2)x^{n-3}$.
Fourth differentiation, $f''''(x) = n(n-1)(n-2)(n-3)x^{n-4}$.

etc., etc.

But this is not the only way of indicating successive differentiations. For, if the original function be

$$y = f(x);$$

differentiating once gives $$\frac{dy}{dx} = f'(x);$$

differentiating twice gives $$\frac{d\left(\dfrac{dy}{dx}\right)}{dx} = f''(x);$$

and this is more conveniently written as $\dfrac{d^2y}{(dx)^2}$, or more usually

$\frac{d^2y}{dx^2}$. Similarly, we may write as the result of differentiating three

times, $\frac{d^3y}{dx^3} = f'''(x)$.

Examples.

Now let us try $\quad y = f(x) = 7x^4 + 3.5x^3 - \frac{1}{2}x^2 + x - 2$

$$\frac{dy}{dx} = f'(x) \quad = 28x^3 + 10.5x^2 - x + 1$$

$$\frac{d^2y}{dx^2} = f''(x) \quad = 84x^2 + 21x - 1$$

$$\frac{d^3y}{dx^3} = f'''(x) \quad = 168x + 21$$

$$\frac{d^4y}{dx^4} = f''''(x) \quad = 168$$

$$\frac{d^5y}{dx^5} = f'''''(x) = 0$$

In a similar manner if $\quad y = \phi(x) = 3x(x^2 - 4)$

$$\phi'(x) = \frac{dy}{dx} \quad = 3[x \times 2x + (x^2 - 4) \times 1] = 3(3x^2 - 4)$$

$$\phi''(x) = \frac{d^2y}{dx^2} = 3 \times 6x = 18x$$

$$\phi'''(x) = \frac{d^3y}{dx^3} = 18$$

$$\phi''''(x) = \frac{d^4y}{dx^4} = 0$$

EXERCISES IV

Find $\dfrac{dy}{dx}$ and $\dfrac{d^2y}{dx^2}$ for the following expressions:

(1) $y = 17x + 12x^2$ 　　　　　　(2) $y = \dfrac{x^2 + a}{x + a}$

(3) $y = 1 + \dfrac{x}{1} + \dfrac{x^2}{1 \times 2} + \dfrac{x^3}{1 \times 2 \times 3} + \dfrac{x^4}{1 \times 2 \times 3 \times 4}$

(4) Find the 2nd and 3rd derivatives in Exercises III, No. 1 to No. 7, and in the Examples given in Chapter VI, No. 1 to No. 7.

Chapter VIII

WHEN TIME VARIES

⸺⸺

Some of the most important problems of the calculus are those where time is the independent variable, and we have to think about the values of some other quantity that varies when the time varies. Some things grow larger as time goes on; some other things grow smaller. The distance that a train has travelled from its starting place goes on ever increasing as time goes on. Trees grow taller as the years go by. Which is growing at the greater rate: a plant 12 inches high which in one month becomes 14 inches high, or a tree 12 feet high which in a year becomes 14 feet high?

In this chapter we are going to make much use of the word *rate*. Nothing to do with birth rate or death rate, though these words suggest so many births or deaths per thousand of the population. When a car whizzes by us, we say: What a terrific rate! When a spendthrift is flinging about his money, we remark that that young man is living at a prodigious rate. What do we mean by *rate*? In both these cases we are making a mental comparison of something that is happening, and the length of time it takes to happen. If the car goes 10 yards per second, a simple bit of mental arithmetic will show us that this is equivalent—while it lasts—to a rate of 600 yards per minute, or over 20 miles per hour.

Now in what sense is it true that a speed of 10 yards per second is the same as 600 yards per minute? Ten yards is not the same as 600 yards, nor is one second the same thing as one minute. What we mean by saying that the *rate* is the same, is this: that the proportion borne between distance passed over and time taken to pass over it, is the same in both cases.

Now try to put some of these ideas into differential notation. Let y in this case stand for money, and let t stand for time.

If you are spending money, and the amount you spend in a short time dt be called dy, the *rate* of spending it will be $\dfrac{dy}{dt}$; or, as regards saving, with a minus sign, as $-\dfrac{dy}{dt}$, because then dy is a *decrement*, not an increment. But money is not a good example for the calculus, because it generally comes and goes by jumps, not by a continuous flow—you may earn \$20,000 a year, but it does not keep running in all day long in a thin stream; it comes in only weekly, or monthly, or quarterly, in lumps: and your expenditure also goes out in sudden payments.

A more apt illustration of the idea of a rate is furnished by the speed of a moving body. From London to Liverpool is 200 miles. If a train leaves London at 7 o'clock, and reaches Liverpool at 11 o'clock, you know that, since it has travelled 200 miles in 4 hours, its average rate must have been 50 miles per hour; because $\frac{200}{4} = \frac{50}{1}$. Here you are really making a mental comparison between the distance passed over and the time taken to pass over it. You are dividing one by the other. If y is the whole distance, and t the whole time, clearly the average rate is $\dfrac{y}{t}$. Now the speed was not actually constant all the way: at starting, and during the slowing up at the end of the journey, the speed was less. Probably at some part, when running downhill, the speed was over 60 miles an hour. If, during any particular element of time dt, the corresponding element of distance passed over was dy, then at that part of the journey the speed was $\dfrac{dy}{dt}$. The *rate* at which one quantity (in the present instance, *distance*) is changing in relation to the other quantity (in this case, *time*) is properly expressed, then, by stating the derivative of one with respect to the other. A *velocity*, scientifically expressed, is the rate at which a very small distance in any given direction is being passed over, and may therefore be written

$$v = \frac{dy}{dt}$$

But if the velocity v is not uniform, then it must be either increasing or else decreasing. The rate at which a velocity is increasing is called the *acceleration*. If a moving body is, at any particular instant, gaining an additional velocity dv in an element of time dt, then the acceleration a at that instant may be written

$$a = \frac{dv}{dt}$$

But since $v = \dfrac{dy}{dt}$,

$$a = \frac{dv}{dt} = \frac{d}{dt}\left(\frac{dy}{dt}\right)$$

which is usually written $\quad a = \dfrac{d^2y}{dt^2};$

or the acceleration is the second derivative of the distance, with respect to time. Acceleration is expressed as a change of velocity in unit time, for instance, as being so many feet per second per second; the notation used being ft/sec^2.

When a railway train has just begun to move, its velocity v is small; but it is rapidly gaining speed—it is being hurried up, or accelerated, by the effort of the engine. So its $\dfrac{d^2y}{dt^2}$ is large. When it has got up its top speed it is no longer being accelerated, so that then $\dfrac{d^2y}{dt^2}$ has fallen to zero. But when it nears its stopping place its speed begins to slow down; may, indeed, slow down very quickly if the brakes are put on, and during this period of *deceleration* or slackening of pace, the value of $\dfrac{dv}{dt}$, that is, of $\dfrac{d^2y}{dt^2}$, will be negative.

To accelerate a mass m requires the continuous application of force. The force necessary to accelerate a mass is proportional to the mass, and it is also proportional to the acceleration which is being imparted. Hence we may write for the force f, the expression

$$f = ma$$

or
$$f = m\frac{dv}{dt}$$

or
$$f = m\frac{d^2y}{dt^2}$$

The product of a mass by the speed at which it is going is called its *momentum*, and is in symbols mv. If we differentiate momentum with respect to time we shall get $\dfrac{d(mv)}{dt}$ for the rate of change of momentum. But, since m is a constant quantity, this may be written $m\dfrac{dv}{dt}$, which we see above is the same as f. That is to say, force may be expressed either as mass times acceleration, or as rate of change of momentum.

Again, if a force is employed to move something (against an equal and opposite counter-force), it does *work*; and the amount of work done is measured by the product of the force into the distance (in its own direction) through which its point of application moves forward. So if a force f moves forward through a length y, the work done (which we may call w) will be

$$w = f \times y$$

where we take f as a constant force. If the force varies at different parts of the range y, then we must find an expression for its value from point to point. If f be the force along the small element of length dy, the amount of work done will be $f \times dy$. But as dy is only an element of length, only an element of work will be done. If we write w for work, then an element of work will be dw; and we have

$$dw = f \times dy$$

which may be written $dw = ma \cdot dy$;

or
$$dw = m\frac{d^2y}{dt^2} \cdot dy$$

or
$$dw = m\frac{dv}{dt} \cdot dy$$

Further, we may transpose the expression and write

$$\frac{dw}{dy} = f$$

This gives us yet a third definition of *force;* that if it is being used to produce a displacement in any direction, the force (in that direction) is equal to the rate at which work is being done per unit of length in that direction. In this last sentence the word *rate* is clearly not used in its time-sense, but in its meaning as ratio or proportion.

Sir Isaac Newton, who was (along with Leibniz) an inventor of the methods of the calculus, regarded all quantities that were varying as *flowing;* and the ratio which we nowadays call the derivative he regarded as the rate of flowing, or the *fluxion* of the quantity in question. He did not use the notation of the dy and dx, and dt (this was due to Leibniz), but had instead a notation of his own. If y was a quantity that varied, or "flowed", then his symbol for its rate of variation (or "fluxion") was $\dot{y}$. If x was the variable, then its fluxion was called $\dot{x}$. The dot over the letter indicated that it had been differentiated. But this notation does not tell us what is the independent variable with respect to which the differentiation has been effected. When we see $\frac{dy}{dt}$ we know that y is to be differentiated with respect to t. If we see $\frac{dy}{dx}$ we know that y is to be differentiated with respect to x. But if we see merely $\dot{y}$, we cannot tell without looking at the context whether this is to mean $\frac{dy}{dx}$ or $\frac{dy}{dt}$ or $\frac{dy}{dz}$, or what is the other variable. So, therefore, this fluxional notation is less informing than the differential notation, and has in consequence largely dropped out of use. But its simplicity gives it an advantage if only we will agree to use it for those cases exclusively where *time* is the independent variable. In that case $\dot{y}$ will mean $\frac{dy}{dt}$ and $\dot{u}$ will mean $\frac{du}{dt}$; and $\ddot{x}$ will mean $\frac{d^2x}{dt^2}$.

Adopting this fluxional notation we may write the mechanical equations considered in the paragraphs above, as follows:

distance	x
velocity	$v = \dot{x}$
acceleration	$a = \dot{v} = \ddot{x}$
force	$f = m\dot{v} = m\ddot{x}$
work	$w = x \times m\ddot{x}$

Examples.

(1) A body moves so that the distance x (in feet), which it travels from a certain point O, is given by the relation

$$x = 0.2t^2 + 10.4$$

where t is the time in seconds elapsed since a certain instant. Find the velocity and acceleration 5 seconds after the body began to move, and also find the corresponding values when the distance covered is 100 feet. Find also the average velocity during the first 10 seconds of its motion. (Suppose distances and motion to the right to be positive.)

Now $x = 0.2t^2 + 10.4$

$$v = \dot{x} = \frac{dx}{dt} = 0.4t; \quad \text{and} \quad a = \ddot{x} = \frac{d^2x}{dt^2} = 0.4 = \text{constant.}$$

When $t = 0$, $x = 10.4$ and $v = 0$. The body started from a point 10.4 feet to the right of the point O; and the time was reckoned from the instant the body started.

When $t = 5$, $v = 0.4 \times 5 = 2$ ft./sec.; $a = 0.4$ ft./sec².

When $x = 100$, $100 = 0.2t^2 + 10.4$,   or $t^2 = 448$,

and $t = 21.17$ sec.; $v = 0.4 \times 21.17 = 8.468$ ft./sec.

When $t = 10$,

distance travelled $= 0.2 \times 10^2 + 10.4 - 10.4 = 20$ ft.

Average velocity $= \frac{20}{10} = 2$ ft./sec.

(It is the same velocity as the velocity at the middle of the interval, $t = 5$; for, the acceleration being constant, the velocity has varied uniformly from zero when $t = 0$ to 4 ft./sec. when $t = 10$.)

(2) In the above problem let us suppose

$$x = 0.2t^2 + 3t + 10.4$$

$$v = \dot{x} = \frac{dx}{dt} = 0.4t + 3; \quad a = \ddot{x} = \frac{d^2x}{dt^2} = 0.4 = \text{constant}.$$

When $t = 0$, $x = 10.4$ and $v = 3$ ft./sec., the time is reckoned from the instant at which the body passed a point 10.4 ft. from the point 0, its velocity being then already 3 ft./sec. To find the time elapsed since it began moving, let $v = 0$; then $0.4t + 3 = 0$, $t = -\frac{3}{.4} = -7.5$ sec. The body began moving 7.5 sec. before time was begun to be observed; 5 seconds after this gives $t = -2.5$ and $v = 0.4 \times -2.5 + 3 = 2$ ft./sec.

When $x = 100$ ft.,

$$100 = 0.2t^2 + 3t + 10.4; \quad \text{or} \quad t^2 + 15t - 448 = 0$$

hence $t = 14.96$ sec., $v = 0.4 \times 14.96 + 3 = 8.98$ ft./sec.

To find the distance travelled during the first 10 seconds of the motion one must know how far the body was from the point 0 when it started.

When $t = -7.5$,

$$x = 0.2 \times (-7.5)^2 - 3 \times 7.5 + 10.4 = -0.85 \text{ ft.}$$

that is 0.85 ft. to the left of the point 0.

Now, when $t = 2.5$,

$$x = 0.2 \times 2.5^2 + 3 \times 2.5 + 10.4 = 19.15$$

So, in 10 seconds, the distance travelled was $19.15 + 0.85 = 20$ ft., and

$$\text{the average velocity} = \tfrac{20}{10} = 2 \text{ ft./sec.}$$

(3) Consider a similar problem when the distance is given by $x = 0.2t^2 - 3t + 10.4$. Then $v = 0.4t - 3$, $a = 0.4 = \text{constant}$. When $t = 0$, $x = 10.4$ as before, and $v = -3$; so that the body was moving in the direction opposite to its motion in the previous

cases. As the acceleration is positive, however, we see that this velocity will decrease as time goes on, until it becomes zero, when $v = 0$ or $0.4t - 3 = 0$; or $t = +7.5$ sec. After this, the velocity becomes positive; and 5 seconds after the body started, $t = 12.5$, and

$$v = 0.4 \times 12.5 - 3 = 2 \text{ ft./sec.}$$

When $x = 100$,

$$100 = 0.2t^2 - 3t + 10.4, \quad \text{or} \quad t^2 - 15t - 448 = 0$$

and $\quad t = 29.96; \ v = 0.4 \times 29.96 - 3 = 8.98$ ft./sec.

When v is zero, $x = 0.2 \times 7.5^2 - 3 \times 7.5 + 10.4 = -0.85$, informing us that the body moves back to 0.85 ft. beyond the point O before it stops. Ten seconds later $t = 17.5$ and

$$x = 0.2 \times 17.5^2 - 3 \times 17.5 + 10.4 = 19.15$$

The distance travelled $= 0.85 + 19.15 = 20.0$, and the average velocity is again 2 ft./sec.

(4) Consider yet another problem of the same sort with $x = 0.2t^3 - 3t^2 + 10.4$; $v = 0.6t^2 - 6t$; $a = 1.2t - 6$. The acceleration is no more constant.

When $t = 0$, $x = 10.4$, $v = 0$, $a = -6$. The body is at rest, but just ready to move with a negative acceleration, that is, to gain a velocity towards the point O.

(5) If we have $x = 0.2t^3 - 3t + 10.4$, then $v = 0.6t^2 - 3$, and $a = 1.2t$.

When $t = 0$, $x = 10.4$; $v = -3$; $a = 0$.

The body is moving towards the point O with a velocity of 3 ft./sec., and just at that instant the velocity is uniform.

We see that the conditions of the motion can always be at once ascertained from the time-distance equation and its first and second derived functions. In the last two cases the mean velocity during the first 10 seconds and the velocity 5 seconds after the start will no more be the same, because the velocity is not increasing uniformly, the acceleration being no longer constant.

(6) The angle θ (in radians) turned through by a wheel is given by $\theta = 3 + 2t - 0.1t^3$, where t is the time in seconds from a certain instant; find the angular velocity ω and the angular accelera-

tion α, (*a*) after 1 second; (*b*) after it has performed one revolution. At what time is it at rest, and how many revolutions has it performed up to that instant?

$$\omega = \dot{\theta} = \frac{d\theta}{dt} = 2 - 0.3t^2, \quad \alpha = \ddot{\theta} = \frac{d^2\theta}{dt^2} = -0.6t$$

When $t = 0$, $\theta = 3$; $\omega = 2$ rad./sec.; $\alpha = 0$.
When $t = 1$, $\omega = 2 - 0.3 = 1.7$ rad./sec.; $\alpha = -0.6$ rad./sec².
This is a retardation; the wheel is slowing down.
After 1 revolution

$$\theta = 2\pi = 3 + 2t - 0.1t^3$$

By solving this equation numerically we can get the value or values of t for which $\theta = 2\pi$; these are about 2.11 and 3.02 (there is a third negative value).

When $t = 2.11$,

$$\theta = 6.28; \ \omega = 2 - 1.34 = 0.66 \text{ rad./sec.}$$

$$\alpha = -1.27 \text{ rad./sec}^2$$

When $t = 3.02$,

$$\theta = 6.28; \ \omega = 2 - 2.74 = -0.74 \text{ rad./sec.}$$

$$\alpha = -1.81 \text{ rad./sec}^2$$

The velocity is reversed. The wheel is evidently at rest between these two instants; it is at rest when $\omega = 0$, that is when $0 = 2 - 0.3t^2$, or when $t = 2.58$ sec., it has performed

$$\frac{\theta}{2\pi} = \frac{3 + 2 \times 2.58 - 0.1 \times 2.58^3}{6.28} = 1.025 \text{ revolutions}$$

EXERCISES V

(1) If $y = a + bt^2 + ct^4$; find $\dfrac{dy}{dt}$ and $\dfrac{d^2y}{dt^2}$.

Ans. $\dfrac{dy}{dt} = 2bt + 4ct^3; \ \dfrac{d^2y}{dt^2} = 2b + 12ct^2$

(2) A body falling freely in space describes in t seconds a space s, in feet, expressed by the equation $s = 16t^2$. Draw a curve showing the relation between s and t. Also determine the velocity of the body at the following times from its being let drop: $t = 2$ seconds; $t = 4.6$ seconds; $t = 0.01$ second.[1]

(3) If $x = at - \frac{1}{2}gt^2$; find $\dot{x}$ and $\ddot{x}$.

(4) If a body moves according to the law

$$s = 12 - 4.5t + 6.2t^2$$

find its velocity when $t = 4$ seconds; s being in feet.

(5) Find the acceleration of the body mentioned in the preceding example. Is the acceleration the same for all values of t?

(6) The angle θ (in radians) turned through by a revolving wheel is connected with the time t (in seconds) that has elapsed since starting, by the law

$$\theta = 2.1 - 3.2t + 4.8t^2$$

Find the angular velocity (in radians per second) of that wheel when $1\frac{1}{2}$ seconds have elapsed. Find also its angular acceleration.

---

1. It is good to be clear on just how derivatives apply to falling bodies, the most familiar instance of accelerated motion. Let t be the time in seconds from the moment a stone is dropped, and s the distance it has fallen. The function relating these two variables is $s = 16t^2$. (It graphs as a neat parabola.) Thus after one second the stone has fallen 16 feet, after two seconds it has dropped $4 \times 16 = 64$ feet, after three seconds, $9 \times 16 = 144$ feet, and so on. The first derivative is $32t$. This gives the instantaneous velocity at which the stone is falling at the end of t seconds. After the first second its velocity is 32 feet per second. After two seconds its velocity is 64 feet per second, and so on.

The second derivative is simply 32. This derivative of a derivative of a function" is the stone's acceleration—the rate at which its velocity is increasing. Physicists call it the "gravitation constant" for bodies falling to the earth's surface.

Later on, in chapters on integration, you will see how integrating the falling stone's first derivative will give the distance traveled by the stone between any two moments as it goes from the start of its fall until the time it is stopped by the ground.—M.G.

(7) A slider moves so that, during the first part of its motion, its distance s in inches from its starting point is given by the expression

$$s = 6.8t^3 - 10.8t; \ t \text{ being in seconds.}$$

Find the expression for the velocity and the acceleration at any time; and hence find the velocity and the acceleration after 3 seconds.

(8) The motion of a rising balloon is such that its height h, in miles, is given at any instant by the expression

$$h = 0.5 + \tfrac{1}{10}\sqrt[3]{t - 125}; \ t \text{ being in seconds.}$$

Find an expression for the velocity and the acceleration at any time. Draw curves to show the variation of height, velocity and acceleration during the first ten minutes of the ascent.

(9) A stone is thrown downwards into water and its depth p in meters at any instant t seconds after reaching the surface of the water is given by the expression

$$p = \frac{4}{4 + t^2} + 0.8t - 1$$

Find an expression for the velocity and the acceleration at any time. Find the velocity and acceleration after 10 seconds.

(10) A body moves in such a way that the space described in the time t from starting is given by $s = t^n$, where n is a constant. Find the value of n when the velocity is doubled from the 5th to the 10th second; find it also when the velocity is numerically equal to the acceleration at the end of the 10th second.

Chapter IX

INTRODUCING A
USEFUL DODGE

———— ⤳⤳⤳ ————

Sometimes one is stumped by finding that the expression to be differentiated is too complicated to tackle directly.

Thus, the equation

$$y = (x^2 + a^2)^{\frac{3}{2}}$$

is awkward to a beginner.

Now the dodge to turn the difficulty is this: Write some symbol, such as u, for the expression $x^2 + a^2$; then the equation becomes

$$y = u^{\frac{3}{2}}$$

which you can easily manage; for

$$\frac{dy}{du} = \frac{3}{2} u^{\frac{1}{2}}$$

Then tackle the expression

$$u = x^2 + a^2$$

and differentiate it with respect to x

$$\frac{du}{dx} = 2x$$

Then all that remains is plain sailing;

for $$\frac{dy}{dx} = \frac{dy}{du} \times \frac{du}{dx}$$

that is,
$$\frac{dy}{dx} = \frac{3}{2}u^{\frac{1}{2}} \times 2x$$
$$= \tfrac{3}{2}(x^2 + a^2)^{\frac{1}{2}} \times 2x$$
$$= 3x(x^2 + a^2)^{\frac{1}{2}}$$

and so the trick is done.[1]

By and by, when you have learned how to deal with sines, and cosines, and exponentials, you will find this dodge of increasing usefulness.

Examples.

Let us practice this dodge on a few examples.

(1) Differentiate $y = \sqrt{a + x}$.

Let $u = a + x$.

$$\frac{du}{dx} = 1; \; y = u^{\frac{1}{2}}; \; \frac{dy}{du} = \tfrac{1}{2}u^{-\frac{1}{2}} = \tfrac{1}{2}(a + x)^{-\frac{1}{2}}$$

$$\frac{dy}{dx} = \frac{dy}{du} \times \frac{du}{dx} = \frac{1}{2\sqrt{a + x}}$$

1. Thompson's "useful dodge" is known today as the "chain rule." It is one of the most useful rules in calculus. The function he gives to illustrate it is called a "composite function" because it involves a "function of a function." The expression inside the parentheses $(x^2 + a^2)$ is called the "inside function." The "outside function" is the exponent $3/2$.

One could try to differentiate by cubing $(x^2 + a^2)$ and taking its square root, or by expanding $(x^2 + a^2)^{\frac{3}{2}}$ by the binomial theorem, but neither of these methods works very well. As Thompson makes clear, the much simpler way is to differentiate the outside function with respect to the inside one, then multiply the result by the derivative of the inside function with respect to x. It is called the chain rule because it can be applied to composite functions with more than one inside function. You simply compute a chain of derivatives, then multiply them together. Modern calculus texts give proof of why the chain rule works.

For a trivial example of how it works, consider three children A, B, and C. A grows twice as fast as B, and B grows three times as fast as C. How much faster is A growing than C? Clearly the answer is $2 \times 3 = 6$ times as fast.—M.G.

(2) Differentiate $y = \dfrac{1}{\sqrt{a + x^2}}$

Let $u = a + x^2$.

$$\frac{du}{dx} = 2x; \; y = u^{-\frac{1}{2}}; \; \frac{dy}{du} = -\tfrac{1}{2}u^{-\frac{3}{2}}$$

$$\frac{dy}{dx} = \frac{dy}{du} \times \frac{du}{dx} = -\frac{x}{\sqrt{(a + x^2)^3}}$$

(3) Differentiate $y = \left(m - nx^{\frac{2}{3}} + \dfrac{p}{x^{\frac{4}{3}}} \right)^a$

Let $u = m - nx^{\frac{2}{3}} + px^{-\frac{4}{3}}$.

$$\frac{du}{dx} = -\tfrac{2}{3}nx^{-\frac{1}{3}} - \tfrac{4}{3}px^{-\frac{7}{3}}$$

$$y = u^a; \; \frac{dy}{du} = au^{a-1}$$

$$\frac{dy}{dx} = \frac{dy}{du} \times \frac{du}{dx} = -a\left(m - nx^{\frac{2}{3}} + \frac{p}{x^{\frac{4}{3}}} \right)^{a-1} \left(\tfrac{2}{3}nx^{-\frac{1}{3}} + \tfrac{4}{3}px^{-\frac{7}{3}} \right)$$

(4) Differentiate $y = \dfrac{1}{\sqrt{x^3 - a^2}}$

Let $u = x^3 - a^2$.

$$\frac{du}{dx} = 3x^2; \; y = u^{-\frac{1}{2}}; \; \frac{dy}{du} = -\tfrac{1}{2}(x^3 - a^2)^{-\frac{3}{2}}$$

$$\frac{dy}{dx} = \frac{dy}{du} \times \frac{du}{dx} = -\frac{3x^2}{2\sqrt{(x^3 - a^2)^3}}$$

(5) Differentiate $y = \sqrt{\dfrac{1 - x}{1 + x}}$

Write this as $y = \dfrac{(1 - x)^{\frac{1}{2}}}{(1 + x)^{\frac{1}{2}}}$

$$\frac{dy}{dx} = \frac{(1+x)^{\frac{1}{2}}\dfrac{d(1-x)^{\frac{1}{2}}}{dx} - (1-x)^{\frac{1}{2}}\dfrac{d(1+x)^{\frac{1}{2}}}{dx}}{1+x}$$

(We may also write $y = (1-x)^{\frac{1}{2}}(1+x)^{-\frac{1}{2}}$ and differentiate as a product.)

Proceeding as in Example (1) above, we get

$$\frac{d(1-x)^{\frac{1}{2}}}{dx} = -\frac{1}{2\sqrt{1-x}}; \quad \text{and} \quad \frac{d(1+x)^{\frac{1}{2}}}{dx} = \frac{1}{2\sqrt{1+x}}$$

Hence $\quad \dfrac{dy}{dx} = -\dfrac{(1+x)^{\frac{1}{2}}}{2(1+x)\sqrt{1-x}} - \dfrac{(1-x)^{\frac{1}{2}}}{2(1+x)\sqrt{1+x}}$

$$= -\frac{1}{2\sqrt{1+x}\,\sqrt{1-x}} - \frac{\sqrt{1-x}}{2\sqrt{(1+x)^3}}$$

or $\quad \dfrac{dy}{dx} = -\dfrac{1}{(1+x)\sqrt{1-x^2}}$

(6) Differentiate $y = \sqrt{\dfrac{x^3}{1+x^2}}$

We may write this

$$y = x^{\frac{3}{2}}(1+x^2)^{-\frac{1}{2}}$$

$$\frac{dy}{dx} = \frac{3}{2}x^{\frac{1}{2}}(1+x^2)^{-\frac{1}{2}} + x^{\frac{3}{2}} \times \frac{d\left[(1+x^2)^{-\frac{1}{2}}\right]}{dx}$$

Differentiating $(1+x^2)^{-\frac{1}{2}}$, as shown in Example (2) above, we get

$$\frac{d\left[(1+x^2)^{-\frac{1}{2}}\right]}{dx} = -\frac{x}{\sqrt{(1+x^2)^3}}$$

so that $\quad \dfrac{dy}{dx} = \dfrac{3\sqrt{x}}{2\sqrt{1+x^2}} - \dfrac{\sqrt{x^5}}{\sqrt{(1+x^2)^3}} = \dfrac{\sqrt{x}(3+x^2)}{2\sqrt{(1+x^2)^3}}$

(7) Differentiate $y = \left(x + \sqrt{x^2+x+a}\right)^3$

Let $u = x + \sqrt{x^2+x+a}$

$$\frac{du}{dx} = 1 + \frac{d\left[(x^2 + x + a)^{\frac{1}{2}}\right]}{dx}$$

$$y = u^3; \quad \text{and} \quad \frac{dy}{du} = 3u^2 = 3\left(x + \sqrt{x^2 + x + a}\right)^2$$

Now let $v = (x^2 + x + a)^{\frac{1}{2}}$ and $w = (x^2 + x + a)$

$$\frac{dw}{dx} = 2x + 1; \; v = w^{\frac{1}{2}}; \; \frac{dv}{dw} = \frac{1}{2}w^{-\frac{1}{2}}$$

$$\frac{dv}{dx} = \frac{dv}{dw} \times \frac{dw}{dx} = \frac{1}{2}(x^2 + x + a)^{-\frac{1}{2}}(2x + 1)$$

Hence $\quad \dfrac{du}{dx} = 1 + \dfrac{2x + 1}{2\sqrt{x^2 + x + a}}$

$$\frac{dy}{dx} = \frac{dy}{du} \times \frac{du}{dx}$$

$$= 3\left(x + \sqrt{x^2 + x + a}\right)^2 \left(1 + \frac{2x + 1}{2\sqrt{x^2 + x + a}}\right)$$

(8) Differentiate $y = \sqrt{\dfrac{a^2 + x^2}{a^2 - x^2}} \; \sqrt[3]{\dfrac{a^2 - x^2}{a^2 + x^2}}$

We get $\quad y = \dfrac{(a^2 + x^2)^{\frac{1}{2}}(a^2 - x^2)^{\frac{1}{3}}}{(a^2 - x^2)^{\frac{1}{2}}(a^2 + x^2)^{\frac{1}{3}}} = (a^2 + x^2)^{\frac{1}{6}} (a^2 - x^2)^{-\frac{1}{6}}$

$$\frac{dy}{dx} = (a^2 + x^2)^{\frac{1}{6}} \frac{d\left[(a^2 - x^2)^{-\frac{1}{6}}\right]}{dx} + \frac{d\left[(a^2 + x^2)^{\frac{1}{6}}\right]}{(a^2 - x^2)^{\frac{1}{6}}dx}$$

Let $u = (a^2 - x^2)^{-\frac{1}{6}}$ and $v = (a^2 - x^2)$

$$u = v^{-\frac{1}{6}}; \; \frac{du}{dv} = -\frac{1}{6}v^{-\frac{7}{6}}; \; \frac{dv}{dx} = -2x$$

$$\frac{du}{dx} = \frac{du}{dv} \times \frac{dv}{dx} = \frac{1}{3}x(a^2 - x^2)^{-\frac{7}{6}}$$

Let $w = (a^2 + x^2)^{\frac{1}{6}}$ and $z = (a^2 + x^2)$

$$w = z^{\frac{1}{6}}; \quad \frac{dw}{dz} = \frac{1}{6}z^{-\frac{5}{6}}; \quad \frac{dz}{dx} = 2x$$

$$\frac{dw}{dx} = \frac{dw}{dz} \times \frac{dz}{dx} = \frac{1}{3}x(a^2 + x^2)^{-\frac{5}{6}}$$

Hence $\quad \dfrac{dy}{dx} = (a^2 + x^2)^{\frac{1}{6}} \dfrac{x}{3(a^2 - x^2)^{\frac{7}{6}}} + \dfrac{x}{3(a^2 - x^2)^{\frac{1}{6}}(a^2 + x^2)^{\frac{5}{6}}}$

or $\quad \dfrac{dy}{dx} = \dfrac{x}{3}\left[\sqrt[6]{\dfrac{a^2 + x^2}{(a^2 - x^2)^7}} + \dfrac{1}{\sqrt[6]{(a^2 - x^2)(a^2 + x^2)^5}} \right]$

(9) Differentiate y'' with respect to y^5.

$$\frac{d(y'')}{d(y^5)} = \frac{ny^{n-1}}{5y^{5-1}} = \frac{n}{5}y^{n-5}$$

(10) Find the first and second derivatives of $y = \dfrac{x}{b}\sqrt{(a - x)x}$.

$$\frac{dy}{dx} = \frac{x}{b}\frac{d\{[(a - x)x]^{\frac{1}{2}}\}}{dx} + \frac{\sqrt{(a - x)x}}{b}$$

Let $u = [(a - x)x]^{\frac{1}{2}}$ and let $w = (a - x)x$; then $u = w^{\frac{1}{2}}$

$$\frac{du}{dw} = \frac{1}{2}w^{-\frac{1}{2}} = \frac{1}{2w^{\frac{1}{2}}} = \frac{1}{2\sqrt{(a - x)x}}$$

$$\frac{dw}{dx} = a - 2x$$

$$\frac{du}{dw} \times \frac{dw}{dx} = \frac{du}{dx} = \frac{a - 2x}{2\sqrt{(a - x)x}}$$

Hence $\quad \dfrac{dy}{dx} = \dfrac{x(a - 2x)}{2b\sqrt{(a - x)x}} + \dfrac{\sqrt{(a - x)x}}{b} = \dfrac{x(3a - 4x)}{2b\sqrt{(a - x)x}}$

Now $\dfrac{d^2y}{dx^2} = \dfrac{2b\sqrt{(a-x)x}(3a-8x) - \dfrac{(3ax-4x^2)b(a-2x)}{\sqrt{(a-x)x}}}{4b^2(a-x)x}$

$$= \dfrac{3a^2 - 12ax + 8x^2}{4b(a-x)\sqrt{(a-x)x}}$$

(We shall need these two last derivatives later on. See Ex. X, No. 11.)

EXERCISES VI

Differentiate the following:

(1) $y = \sqrt{x^2 + 1}$ (2) $y = \sqrt{x^2 + a^2}$

(3) $y = \dfrac{1}{\sqrt{a+x}}$ (4) $y = \dfrac{a}{\sqrt{a-x^2}}$ (5) $y = \dfrac{\sqrt{x^2 - a^2}}{x^2}$

(6) $y = \dfrac{\sqrt[3]{x^4 + a}}{\sqrt[2]{x^3 + a}}$ (7) $y = \dfrac{a^2 + x^2}{(a+x)^2}$

(8) Differentiate y^5 with respect to y^2

(9) Differentiate $y = \dfrac{\sqrt{1 - \theta^2}}{1 - \theta}$

The process can be extended to three or more derivatives, so that $\dfrac{dy}{dx} = \dfrac{dy}{dz} \times \dfrac{dz}{dv} \times \dfrac{dv}{dx}$

Examples.

(1) If $z = 3x^4$; $v = \dfrac{7}{z^2}$; $y = \sqrt{1 + v}$, find $\dfrac{dy}{dx}$

We have $\dfrac{dy}{dv} = \dfrac{1}{2\sqrt{1+v}}$; $\dfrac{dv}{dz} = -\dfrac{14}{z^3}$; $\dfrac{dz}{dx} = 12x^3$

$$\dfrac{dy}{dx} = -\dfrac{168x^3}{\left(2\sqrt{1+v}\right)z^3} = -\dfrac{28}{3x^5\sqrt{9x^8 + 7}}$$

(2) If $t = \dfrac{1}{5\sqrt{\theta}}$; $x = t^3 + \dfrac{t}{2}$; $v = \dfrac{7x^2}{\sqrt[3]{x-1}}$, find $\dfrac{dv}{d\theta}$

$$\frac{dv}{dx} = \frac{7x(5x-6)}{3\sqrt[3]{(x-1)^4}}; \frac{dx}{dt} = 3t^2 + \tfrac{1}{2}; \frac{dt}{d\theta} = -\frac{1}{10\sqrt{\theta^3}}$$

Hence $\qquad \dfrac{dv}{d\theta} = -\dfrac{7x(5x-6)\left(3t^2 + \tfrac{1}{2}\right)}{30\sqrt[3]{(x-1)^4}\sqrt{\theta^3}}$

an expression in which x must be replaced by its value, and t by its value in terms of θ.

(3) If $\quad \theta = \dfrac{3a^2x}{\sqrt{x^3}}$; $\quad \omega = \dfrac{\sqrt{1-\theta^2}}{1+\theta}$ and $\quad \phi = \sqrt{3} - \dfrac{1}{\omega\sqrt{2}}$,

find $\dfrac{d\phi}{dx}$.

We get $\theta = 3a^2x^{-\frac{1}{2}}$; $\omega = \sqrt{\dfrac{1-\theta}{1+\theta}}$; and $\phi = \sqrt{3} - \dfrac{1}{\sqrt{2}}\omega^{-1}$

$$\frac{d\theta}{dx} = -\frac{3a^2}{2\sqrt{x^3}}; \frac{d\omega}{d\theta} = -\frac{1}{(1+\theta)\sqrt{1-\theta^2}}$$

(see example 5, Capter IX); and

$$\frac{d\phi}{d\omega} = \frac{1}{\sqrt{2}\cdot\omega^2}$$

So that $\quad \dfrac{d\phi}{dx} = \dfrac{1}{\sqrt{2}\times\omega^2} \times \dfrac{1}{(1+\theta)\sqrt{1-\theta^2}} \times \dfrac{3a^2}{2\sqrt{x^3}}$

Replace now first ω, then θ by its value.

EXERCISES VII

You can now successfully try the following.

(1) If $u = \tfrac{1}{2}x^3$; $v = 3(u + u^2)$; and $w = \dfrac{1}{v^2}$, find $\dfrac{dw}{dx}$

(2) If $y = 3x^2 + \sqrt{2}$; $z = \sqrt{1+y}$; and $v = \dfrac{1}{\sqrt{3}+4z}$, find $\dfrac{dv}{dx}$

(3) If $y = \dfrac{x^3}{\sqrt{3}}$; $z = (1+y)^2$; and $u = \dfrac{1}{\sqrt{1+z}}$, find $\dfrac{du}{dx}$

The following exercises are placed here for reasons of space and because their solution depends upon the dodge explained in the foregoing chapter, but they should not be attempted until Chapters XIV and XV have been read.

(4) If $y = 2a^3 \log_e u - u\left(5a^2 - 2au + \tfrac{1}{3}u^2\right)$, and $u = a + x$,

show that $\qquad\qquad \dfrac{dy}{dx} = \dfrac{x^2(a-x)}{a+x}$

(5) For the curve $x = a(\theta - \sin\theta)$, $y = a(1 - \cos\theta)$, find $\dfrac{dx}{d\theta}$ and $\dfrac{dy}{d\theta}$; hence deduce the value of $\dfrac{dy}{dx}$

(6) Find $\dfrac{dx}{d\theta}$ and $\dfrac{dy}{d\theta}$ for the curve $x = a\cos^3\theta$, $y = a\sin^3\theta$; hence obtain $\dfrac{dy}{dx}$

(7) Given that $y = \log_e \sin(x^2 - a^2)$, find $\dfrac{dy}{dx}$ in its simplest form.

(8) If $u = x + y$ and $4x = 2u - \log_e(2u - 1)$, show that

$$\dfrac{dy}{dx} = \dfrac{x+y}{x+y-1}$$

Chapter X

GEOMETRICAL MEANING
OF DIFFERENTIATION

It is useful to consider what geometrical meaning can be given to the derivative.

In the first place, any function of x, such, for example, as x^2, or $\sqrt{x}$, or $ax + b$, can be plotted as a curve; and nowadays every schoolboy is familiar with the process of curve plotting.

Let PQR, in Fig. 7, be a portion of a curve plotted with respect to the axes of coordinates OX and OY. Consider any point Q on this curve, where the abscissa of the point is x and its ordinate is y. Now observe how y changes when x is varied. If x is made to increase by a small increment dx, to the right, it will be observed that y also (in *this* particular curve) increases by a small increment dy (because this particular curve happens to be an *ascending* curve). Then the ratio of dy to dx is a measure of the degree to which the curve is sloping up between the two points Q and T. As a matter of fact, it can be seen on the figure that the curve between Q and T has many different slopes, so that we cannot very well speak of the slope of the curve between Q and T. If, however, Q and T are so near each other that the small portion QT of the curve is practically straight, then it is true to say that

the ratio $\dfrac{dy}{dx}$ is the slope of the curve

along QT. The straight line QT produced on either side touches the curve along the portion QT only, and if this portion is infinitely small, the straight

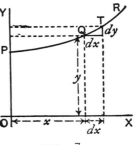

FIG. 7.

103

line will touch the curve at practically one point only, and be therefore a *tangent* to the curve.

This tangent to the curve has evidently the same slope as QT, so that $\dfrac{dy}{dx}$ is the slope of the tangent to the curve at the point Q for which the value of $\dfrac{dy}{dx}$ is found.

We have seen that the short expression "the slope of a curve" has no precise meaning, because a curve has so many slopes—in fact, every small portion of a curve has a different slope. "The slope of a curve *at a point*" is, however, a perfectly defined thing; it is the slope of a very small portion of the curve situated just at that point; and we have seen that this is the same as "the slope of the tangent to the curve at that point".

Observe that dx is a short step to the right, and dy the corresponding short step upwards. These steps must be considered as short as possible—in fact infinitely short,—though in diagrams we have to represent them by bits that are not infinitesimally small, otherwise they could not be seen.

We shall hereafter make considerable use of this circumstance that $\dfrac{dy}{dx}$ *represents the slope of the tangent to the curve at any point.*

If a curve is sloping up at 45° at a particular point, as in Fig. 8, dy and dx will be equal, and the value of $\dfrac{dy}{dx} = 1$.

If the curve slopes up steeper than 45° (Fig. 9), $\dfrac{dy}{dx}$ will be greater than 1.

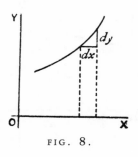

FIG. 8.

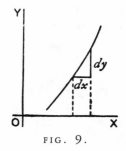

FIG. 9.

If the curve slopes up very gently, as in Fig. 10, $\frac{dy}{dx}$ will be a fraction smaller than 1.

For a horizontal line, or a horizontal place in a curve, $dy = 0$, and therefore $\frac{dy}{dx} = 0$.

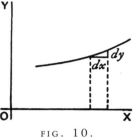

FIG. 10.

If a curve slopes *downward*, as in Fig. 11, dy will be a step down, and must therefore be reckoned of negative value; hence $\frac{dy}{dx}$ will have negative sign also.

If the "curve" happens to be a straight line, like that in Fig. 12, the value of $\frac{dy}{dx}$ will be the same at all points along it. In other words its *slope* is constant.

If a curve is one that turns more upwards as it goes along to the right, the values of $\frac{dy}{dx}$ will become greater and greater with the increasing steepness, as in Fig. 13.

If a curve is one that gets flatter and flatter as it goes along, the values of $\frac{dy}{dx}$ will become smaller and smaller as the flatter part is reached, as in Fig. 14.

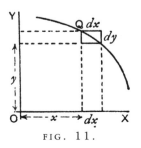

FIG. 11.

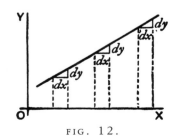

FIG. 12.

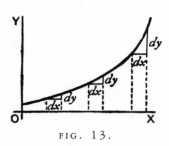

FIG. 13.

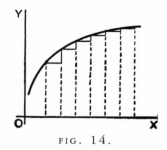

FIG. 14.

If a curve first descends, and then goes up again, as in Fig. 15, presenting a concavity upwards, then clearly $\dfrac{dy}{dx}$ will first be negative, with diminishing values as the curve flattens, then will be zero at the point where the bottom of the trough of the curve is reached; and from this point onward $\dfrac{dy}{dx}$ will have positive values that go on increasing. In such a case y is said to pass through a *local minimum*. This value of y is not necessarily the smallest value of y, it is that value of y corresponding to the bottom of the trough; for instance, in Fig. 28, the value of y corresponding to the bottom of the trough is 1, while y takes elsewhere values which are smaller than this. The characteristic of a local minimum is that y must increase *on either side* of it.

N.B.—For the particular value of x that makes y *a minimum*, the value of $\dfrac{dy}{dx} = 0$.

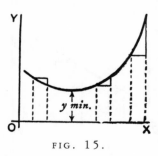

FIG. 15.

If a curve first ascends and then descends, the values of $\dfrac{dy}{dx}$ will be positive at first; then zero, as the summit is reached; then negative, as the curve slopes downwards, as in Fig. 16.

In this case y is said to pass through a *local maximum*, but this value of y is

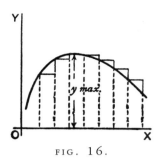

FIG. 16.

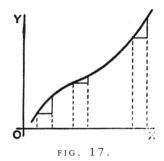

FIG. 17.

not necessarily the greatest value of y. In Fig. 28, the local maximum of y is $2\frac{1}{3}$, but this is by no means the greatest value y can have at some other point of the curve.

N.B.—For the particular value of x that makes y *a maximum,* the value of $\dfrac{dy}{dx} = 0$.

If a curve has the particular form of Fig. 17, the values of $\dfrac{dy}{dx}$ will always be positive; but there will be one particular place where the slope is least steep, where the value of $\dfrac{dy}{dx}$ will be a minimum; that is, less than it is at any other part of the curve.

If a curve has the form of Fig. 18, the value of $\dfrac{dy}{dx}$ will be negative in the upper part, and positive in the lower part; while at the nose of the curve where it becomes actually perpendicular, the value of $\dfrac{dy}{dx}$ will be infinitely great.

Now that we understand that $\dfrac{dy}{dx}$ measures the steepness of a curve at any point, let us turn to some of the equations which we have already learned how to differentiate.

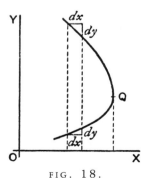

FIG. 18.

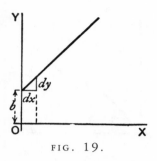

FIG. 19.

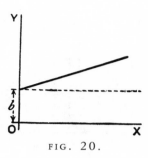

FIG. 20.

(1) As the simplest case take this:

$$y = x + b$$

It is plotted out in Fig. 19, using equal scales for x and y. If we put $x = 0$, then the corresponding ordinate will be $y = b$; that is to say, the "curve" crosses the y-axis at the height b. From here it ascends at $45°$; for whatever values we give to x to the right, we have an equal y to ascend. The line has a gradient of 1 in 1.

Now differentiate $y = x + b$, by the rules we have already learned and we get $\dfrac{dy}{dx} = 1$.

The slope of the line is such that for every little step dx to the right, we go an equal little step dy upward. And this slope is constant—always the same slope.

(2) Take another case:

$$y = ax + b.$$

We know that this curve, like the preceding one, will start from a height b on the y-axis. But before we draw the curve, let us find its slope by differentiating; which gives us $\dfrac{dy}{dx} = a$. The slope will be constant, at an angle, the tangent of which is here called a. Let us assign to a some numerical value—say $\frac{1}{3}$. Then we must give it such a slope that it ascends 1 in 3; or dx will be 3 times as great as dy; as magnified in Fig. 21. So, draw the line in Fig. 20 at this slope.

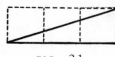

FIG. 21.

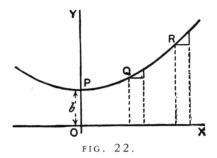

FIG. 22.

(3) Now for a slightly harder case. Let

$$y = ax^2 + b$$

Again the curve will start on the y-axis at a height b above the origin.

Now differentiate. [If you have forgotten, turn back to Chapter V, or, rather, *don't* turn back, but think out the differentiation.]

$$\frac{dy}{dx} = 2ax$$

This shows that the steepness will not be constant: it increases as x increases. At the starting point P, where $x = 0$, the curve (Fig. 22) has no steepness—that is, it is level. On the left of the origin, where x has negative values, $\frac{dy}{dx}$ will also have negative values, or will descend from left to right, as in the Figure.

Let us illustrate this by working out a particular instance. Taking the equation

$$y = \tfrac{1}{4}x^2 + 3$$

and differentiating it, we get

$$\frac{dy}{dx} = \tfrac{1}{2}x$$

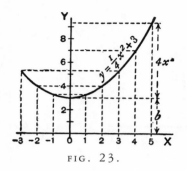

FIG. 23.

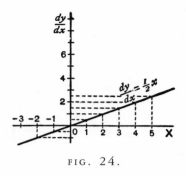

FIG. 24.

Now assign a few successive values, say from 0 to 5, to *x*; and calculate the corresponding values of *y* by the first equation; and of $\dfrac{dy}{dx}$ from the second equation. Tabulating results, we have:

x	0	1	2	3	4	5
y	3	$3\frac{1}{4}$	4	$5\frac{1}{4}$	7	$9\frac{1}{4}$
$\dfrac{dy}{dx}$	0	$\frac{1}{2}$	1	$1\frac{1}{2}$	2	$2\frac{1}{2}$

Then plot them out in two curves, Figs. 23 and 24; in Fig. 23 plotting the values of *y* against those of *x*, and in Fig. 24 those of $\dfrac{dy}{dx}$ against those of *x*. For any assigned value of *x*, the *height* of the ordinate in the second curve is proportional to the *slope* of the first curve.

If a curve comes to a sudden *cusp,* as in Fig. 25, the slope at that point suddenly changes from a slope upward

to a slope downward. In that case $\dfrac{dy}{dx}$

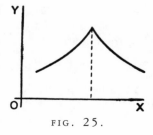

FIG. 25.

will clearly undergo an abrupt change from a positive to a negative value.[1]

The following examples show further applications of the principles just explained.

(4) Find the slope of the tangent to the curve $y = \dfrac{1}{2x} + 3$ at the point where $x = -1$. Find the angle which this tangent makes with the curve $y = 2x^2 + 2$.

The slope of the tangent is the slope of the curve at the point where they touch one another; that is, it is the $\dfrac{dy}{dx}$ of the curve for that point. Here $\dfrac{dy}{dx} = -\dfrac{1}{2x^2}$ and for $x = -1$, $\dfrac{dy}{dx} = -\dfrac{1}{2}$, which is the slope of the tangent and of the curve at that point. The tangent, being a straight line, has for equation $y = ax + b$, and its slope is $\dfrac{dy}{dx} = a$, hence $a = -\dfrac{1}{2}$. Also if $x = -1$, $y = \dfrac{1}{2(-1)} + 3 = 2\frac{1}{2}$; and as the tangent passes by this point, the coordinates of the point must satisfy the equation of the tangent, namely

$$y = -\frac{1}{2}x + b$$

so that $2\frac{1}{2} = -\dfrac{1}{2} \times (-1) + b$ and $b = 2$; the equation of the tangent is therefore $y = -\dfrac{1}{2}x + 2$

---

1. A cusp is the sharp point on a curve where the curve abruptly changes its direction by $180°$. If it changes by some other angle the point is called a "corner." A cusp or corner can, of course, point north as well as south, east, or west, or in any other direction. The tangent at the cusp shown in Figure 25 is vertical. The curve approaches the tangent from one side, leaves it on the other side. At the cusp point there is no derivative. A typical example is the curve for $y = (x - 4)^{\frac{2}{3}}$. Its cusp points south at $x = 4$.

Corners can occur on graphs of continuous functions that are made up of straight lines. For example, the absolute value function, defined to be x if $x \geq 0$ and $-x$ if $x < 0$, has a corner at $x = 0$.—M.G.

Now, when two curves meet, the intersection being a point common to both curves, its coordinates must satisfy the equation of each one of the two curves; that is, it must be a solution of the system of simultaneous equations formed by coupling together the equations of the curves. Here the curves meet one another at points given by the solution of

$$\begin{cases} y = 2x^2 + 2 \\ y = -\tfrac{1}{2}x + 2 \end{cases} \quad \text{or} \quad 2x^2 + 2 = -\tfrac{1}{2}x + 2$$

that is,

$$x\left(2x + \tfrac{1}{2}\right) = 0$$

This equation has for its solutions $x = 0$ and $x = -\tfrac{1}{4}$. The slope of the curve $y = 2x^2 + 2$ at any point is

$$\frac{dy}{dx} = 4x$$

For the point where $x = 0$, this slope is zero; the curve is horizontal. For the point where

$$x = -\frac{1}{4}, \quad \frac{dy}{dx} = -1$$

hence the curve at that point slopes downwards to the right at such an angle θ with the horizontal that $\tan\theta = 1$; that is, at $45°$ to the horizontal.

The slope of the straight line is $-\tfrac{1}{2}$; that is, it slopes downwards to the right and makes with the horizontal an angle ϕ such that $\tan\phi = \tfrac{1}{2}$; that is, an angle of $26°\ 34'$. It follows that at the first point the curve cuts the straight line at an angle of $26°\ 34'$, while at the second it cuts it at an angle of $45° - 26°\ 34' = 18°\ 26'$.

(5) A straight line is to be drawn, through a point whose co-ordinates are $x = 2$, $y = -1$, as tangent to the curve

$$y = x^2 - 5x + 6$$

Find the coordinates of the point of contact.

The slope of the tangent must be the same as the $\dfrac{dy}{dx}$ of the curve; that is, $2x - 5$.

The equation of the straight line is $y = ax + b$, and as it is satisfied for the values $x = 2$, $y = -1$, then $-1 = a \times 2 + b$; also, its

$$\frac{dy}{dx} = a = 2x - 5$$

The x and the y of the point of contact must also satisfy both the equation of the tangent and the equation of the curve.

We have then

$$\left\{ \begin{array}{ll} y = x^2 - 5x + 6, & \dotfill(\text{i}) \\[6pt] y = ax + b, & \dotfill(\text{ii}) \\[6pt] -1 = 2a + b, & \dotfill(\text{iii}) \\[6pt] a = 2x - 5, & \dotfill(\text{iv}) \end{array} \right.$$

four equations in a, b, x, y.

Equations (i) and (ii) give $x^2 - 5x + 6 = ax + b$.

Replacing a and b by their value in this, we get

$$x^2 - 5x + 6 = (2x - 5)x - 1 - 2(2x - 5)$$

which simplifies to $x^2 - 4x + 3 = 0$, the solutions of which are: $x = 3$ and $x = 1$. Replacing in (i), we get $y = 0$ and $y = 2$ respectively; the two points of contact are then $x = 1$, $y = 2$; and $x = 3$, $y = 0$.

Note.—In all exercises dealing with curves, students will find it extremely instructive to verify the deductions obtained by actually plotting the curves.

EXERCISES VIII

(1) Plot the curve $y = \frac{3}{4}x^2 - 5$, using a scale of millimeters. Measure at points corresponding to different values of x, the angle of its slope.

Find, by differentiating the equation, the expression for slope; and see from your calculator whether this agrees with the measured angle.

(2) Find what will be the slope of the curve

$$y = 0.12x^3 - 2$$

at the point where $x = 2$.

(3) If $y = (x - a)(x - b)$, show that at the particular point of the curve where $\dfrac{dy}{dx} = 0$, x will have the value $\frac{1}{2}(a + b)$.

(4) Find the $\dfrac{dy}{dx}$ of the equation $y = x^3 + 3x$; and calculate the numerical values of $\dfrac{dy}{dx}$ for the points corresponding to $x = 0$, $x = \frac{1}{2}$, $x = 1$, $x = 2$.

(5) In the curve to which the equation is $x^2 + y^2 = 4$, find the value of x at those points where the slope $= 1$.

(6) Find the slope, at any point, of the curve whose equation is $\dfrac{x^2}{3^2} + \dfrac{y^2}{2^2} = 1$; and give the numerical value of the slope at the place where $x = 0$, and at that where $x = 1$.

(7) The equation of a tangent to the curve $y = 5 - 2x + 0.5x^3$, being of the form $y = mx + n$, where m and n are constants, find the value of m and n if the point where the tangent touches the curve has $x = 2$ for abscissa.

(8) At what angle do the two curves

$$y = 3.5x^2 + 2 \quad \text{and} \quad y = x^2 - 5x + 9.5$$

cut one another?

(9) Tangents to the curve $y = \pm\sqrt{25 - x^2}$ are drawn at points for which $x = 3$ and $x = 4$, the value of y being positive. Find the coordinates of the point of intersection of the tangents and their mutual inclination.

(10) A straight line $y = 2x - b$ touches a curve $y = 3x^2 + 2$ at one point. What are the coordinates of the point of contact, and what is the value of b?

POSTSCRIPT

If the graph of a continuous function crosses the x axis at two points a and b, and is differentiable for the closed interval be-

tween *a* and *b*, there must be at least
one point on the curve between *a* and
b where the tangent is horizontal and
the derivative is zero. This is called
Rolle's (pronounced Roll's) theorem after French mathematician Michel Rolle
(1652-1719).

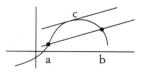

F I G .   2 5 a .   Lagrange's
mean value theorem.

Rolle's theorem is a special case of what
is called Lagrange's mean value theorem, after French mathematician Joseph Louis Lagrange (1736-1813). It states that a
straight line between points *a* and *b* on the curve for a continuous
differentiable function is parallel to the tangent of at least one
point *c* on the closed interval between *a* and *b*. (See Figure 25a)

Applied to the velocity of a car that goes at an average speed of,
say, 45 miles per hour from A to B, no matter how many times
it alters its speed along the way (including even stops), there will
be at least one moment when the car's instantaneous speed will
be exactly 45 miles per hour.

Both theorems are intuitively obvious, yet they underlie many
important, more complicated calculus theorems. For example,
the mean value theorem is the basis for calculating antideriva-
tives.

The Lagrange mean value theorem further generalizes to what is
called Cauchy's mean value theorem after Augustin Louis Cauchy
(1789-1857), a French mathematician. It concerns two continuous functions that are differentiable in a closed interval. You will
learn about it in more advanced calculus textbooks.—M.G.

Chapter XI

MAXIMA AND MINIMA

⌒∞∞∞⌒

A quantity which varies continuously is said to pass by (or through) a local maximum or minimum value when, in the course of its variation, the immediately preceding and following values are *both* smaller or greater, respectively, than the value referred to. An infinitely great value is therefore not a maximum value.

One of the principal uses of the process of differentiating is to find out under what conditions the value of the thing differentiated becomes a maximum or a minimum. This is often exceedingly important in engineering and economic questions, where it is most desirable to know what conditions will make the cost of working a minimum, or will make the efficiency a maximum.

Now, to begin with a concrete case, let us take the equation

$$y = x^2 - 4x + 7$$

By assigning a number of successive values to x, and finding the corresponding values of y, we can readily see that the equation represents a curve with a minimum.

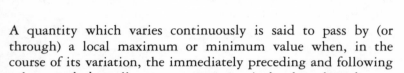

x	0	1	2	3	4	5
y	7	4	3	4	7	12

These values are plotted in Fig. 26, which shows that y has apparently a minimum value of 3, when x is made equal to 2. But are you sure that the minimum occurs at 2, and not at $2\frac{1}{4}$ or at $1\frac{3}{4}$?

Of course it would be possible with any algebraic expression to work out a lot of values, and in this way arrive gradually at the particular value that may be a maximum or a minimum.

Here is another example: Let

$$y = 3x - x^2$$

Calculate a few values thus:

x	-1	0	1	2	3	4	5
y	-4	0	2	2	0	-4	-10

Plot these values as in Fig. 27.

It will be evident that there will be a maximum somewhere between $x = 1$ and $x = 2$; and the thing *looks* as if the maximum value of y ought to be about $2\frac{1}{4}$. Try some intermediate values. If $x = 1\frac{1}{4}$, $y = 2.1875$; if $x = 1\frac{1}{2}$, $y = 2.25$; if $x = 1.6$, $y = 2.24$. How can we be sure that 2.25 is the real maximum, or that it occurs exactly when $x = 1\frac{1}{2}$?

Now it may sound like juggling to be assured that there is a way by which one can arrive straight at a maximum (or minimum) value without making a lot of preliminary trials or guesses. And that way depends on differentiating. Look back to Chapter X for the remarks about Figs. 15 and 16, and you will see that whenever a curve gets either to its maximum or to its

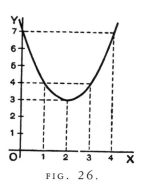

FIG. 26.

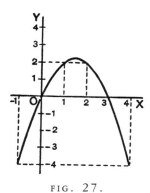

FIG. 27.

minimum height, at that point its $\dfrac{dy}{dx} = 0$. Now this gives us the

clue to the dodge that is wanted. When there is put before you an equation, and you want to find that value of x that will make its y a minimum (or a maximum), *first differentiate it,* and having

done so, write its $\dfrac{dy}{dx}$ as *equal to zero,* and then solve for x. Put this

particular value of x into the original equation, and you will then get the required value of y. This process is commonly called "equating to zero".

To see how simply it works, take the example with which this chapter opens, namely,

$$y = x^2 - 4x + 7$$

Differentiating, we get:

$$\frac{dy}{dx} = 2x - 4$$

Now equate this to zero, thus:

$$2x - 4 = 0$$

Solving this equation for x, we get:

$$2x = 4, \quad x = 2$$

Now, we know that the maximum (or minimum) will occur exactly when $x = 2$.

Putting the value $x = 2$ into the original equation, we get

$$y = 2^2 - (4 \times 2) + 7$$
$$= 4 - 8 + 7 = 3$$

Now look back at Fig. 26, and you will see that the minimum occurs when $x = 2$, and that this minimum of $y = 3$.

Try the second example (Fig. 27), which is

$$y = 3x - x^2$$

Differentiating, $$\frac{dy}{dx} = 3 - 2x$$

Equating to zero, $\quad 3 - 2x = 0$

whence $\qquad\qquad x = 1\tfrac{1}{2}$

and putting this value of x into the original equation, we find:

$$y = 4\tfrac{1}{2} - \left(1\tfrac{1}{2} \times 1\tfrac{1}{2}\right)$$
$$y = 2\tfrac{1}{4}$$

This gives us exactly the information as to which the method of trying a lot of values left us uncertain.

Now, before we go on to any further cases, we have two remarks to make. When you are told to equate $\frac{dy}{dx}$ to zero, you feel at first (that is if you have any wits of your own) a kind of resentment, because you know that $\frac{dy}{dx}$ has all sorts of different values at different parts of the curve, according to whether it is sloping up or down. So, when you are suddenly told to write

$$\frac{dy}{dx} = 0$$

you resent it, and feel inclined to say that it can't be true. Now you will have to understand the essential difference between "an equation", and "an equation of condition". Ordinarily you are dealing with equations that are true in themselves; but, on occasions, of which the present are examples, you have to write down equations that are not necessarily true, but are only true if certain conditions are to be fulfilled; and you write them down in order, by solving them, to find the conditions which make them true. Now we want to find the particular value that x has when the curve is neither sloping up nor sloping down, that is, at the particular place where $\dfrac{dy}{dx} = 0$. So, writing $\dfrac{dy}{dx} = 0$ does *not* mean that it always is $= 0$; but you write it down *as a condition* in order to see how much x will come out if $\dfrac{dy}{dx}$ is to be zero.

The second remark is one which (if you have any wits of your own) you will probably have already made: namely, that this much-belauded process of equating to zero entirely fails to tell you whether the x that you thereby find is going to give you a *maximum* value of y or a *minimum* value of y. Quite so. It does not of itself discriminate; it finds for you the right value of x but leaves you to find out for yourselves whether the corresponding y is a maximum or a minimum. Of course, if you have plotted the curve, you know already which it will be.

For instance, take the equation:

$$y = 4x + \frac{1}{x}$$

Without stopping to think what curve it corresponds to, differentiate it, and equate to zero:

$$\frac{dy}{dx} = 4 - x^{-2} = 4 - \frac{1}{x^2} = 0$$

whence $\qquad\qquad x = \tfrac{1}{2} \quad \text{or} \quad x = -\tfrac{1}{2}$

and, inserting these values,

$$y = 4 \quad \text{or} \quad y = -4$$

Each will be either a maximum or else a minimum. But which? You will hereafter be told a way, depending upon a second differentiation (see Chap. XII). But at present it is enough if you will simply try two other values of x differing a little from the one found, one larger and one smaller, and see whether with these altered values the corresponding values of y are less or greater than that already found.

Try another simple problem in maxima and minima. Suppose you were asked to divide any number into two parts, such that the product was a maximum? How would you set about it if you did not know the trick of equating to zero? I suppose you could worry it out by the rule of try, try, try again. Let 60 be the number. You can try cutting it into two parts, and multiplying them together. Thus, 50 times 10 is 500; 52 times 8 is 416; 40 times 20 is 800; 45 times 15 is 675; 30 times 30 is 900. This looks like a maximum: try varying it. 31 times 29 is 899, which is not so good; and 32 times 28 is 896, which is worse. So it seems that the biggest product will be got by dividing into two halves.

Now see what the calculus tells you. Let the number to be cut into two parts be called n. Then if x is one part, the other will be $n - x$, and the product will be $x(n - x)$ or $nx - x^2$. So we write $y = nx - x^2$. Now differentiate and equate to zero;

$$\frac{dy}{dx} = n - 2x = 0$$

Solving for x, we get $\quad \dfrac{n}{2} = x$

So now we *know* that whatever number n may be, we must divide it into two equal parts if the product of the parts is to be a maximum; and the value of that maximum product will always be $= \frac{1}{4}n^2$.

This is a very useful rule, and applies to any number of factors, so that if $m + n + p =$ a constant number, $m \times n \times p$ is a maximum when $m = n = p$.

Test Case.
Let us at once apply our knowledge to a case that we can test.

Let $\qquad\qquad y = x^2 - x$

and let us find whether this function has a maximum or minimum; and if so, test whether it is a maximum or a minimum.

Differentiating, we get

$$\frac{dy}{dx} = 2x - 1$$

Equating to zero, we get

$$2x - 1 = 0$$

whence $$2x = 1$$

or $$x = \tfrac{1}{2}$$

That is to say, when x is made $= \tfrac{1}{2}$, the corresponding value of y will be either a maximum or a minimum. Accordingly, putting $x = \tfrac{1}{2}$ in the original equation, we get

$$y = \left(\tfrac{1}{2}\right)^2 - \tfrac{1}{2}$$

or $$y = -\tfrac{1}{4}$$

Is this a maximum or a minimum? To test it, try putting x a little bigger than $\tfrac{1}{2}$—say, make $x = 0.6$. Then

$$y = (0.6)^2 - 0.6 = 0.36 - 0.6 = -0.24$$

Also try a value of x a little smaller than $\tfrac{1}{2}$—say, $x = 0.4$.

Then $$y = (0.4)^2 - 0.4 = 0.16 - 0.4 = -0.24$$

Both values are larger than -0.25, showing that $y = -0.25$ is a *minimum*.

Plot the curve for yourself, and verify the calculation.

Further Examples.

A most interesting example is afforded by a curve that has both a maximum and a minimum. Its equation is:

$$y = \tfrac{1}{3}x^3 - 2x^2 + 3x + 1$$

Now $$\frac{dy}{dx} = x^2 - 4x + 3$$

Equating to zero, we get the quadratic,

$$x^2 - 4x + 3 = 0$$

and solving the quadratic gives us *two* roots, viz.

$$\begin{cases} x = 3 \\ x = 1 \end{cases}$$

Now, when $x = 3$, $y = 1$; and when $x = 1$, $y = 2\frac{1}{3}$. The first of these is a minimum, the second a maximum.[1]

The curve itself may be plotted (as in Fig. 28) from the values calculated, as below, from the original equation.

x	-1	0	1	2	3	4	5	6
y	$-4\frac{1}{3}$	1	$2\frac{1}{3}$	$1\frac{2}{3}$	1	$2\frac{1}{3}$	$7\frac{2}{3}$	19

A further exercise in maxima and minima is afforded by the following example:

The equation of a circle of radius r, having its center C at the point whose coordinates are $x = a$, $y = b$, as depicted in Fig. 29, is:

$$(y - b)^2 + (x - a)^2 = r^2$$

This may be transformed into

$$y = \sqrt{r^2 - (x - a)^2} + b$$

(where the square root may be either positive or negative).

---

1. The maximum point on the curve in Figure 28 is called a "local" maximum because the curve obviously has higher points later on. Similarly, its minimum point is a "local" minimum because there are lower points earlier on the curve. If maximum and minimum points are the highest and lowest points on a curve, as in Figures 26 and 27, they are called "absolute" maxima and minima. In Figure 28 the curve has no absolute maximum or minimum points because it goes to infinity at both ends.

In Figure 28 the point on the curve at $x = 2$ is called an "inflection" point. This is a point at which the curve is concave upward on one side and concave downward on the other. Put another way, it is the point at which a tangent to the line, as you move left to right, stops rotating in one direction and starts to rotate the other way. When a curve's second derivative is positive, the curve is concave upward (like a smile), and when the second derivative is negative, the curve is concave downward (like a frown). Calculus texts often distinguish the two curves by saying that one "holds water" and the other does not.

Sometimes a curve has an inflection point at a point where $\frac{dy}{dx} = 0$. In this case, the point is neither a maximum or minimum. For example, this is true for $y = x^3$ at $x = 0$. So after we equate the derivative to zero, we need to check points on both sides of x to determine if we've found a maximum, a minimum, or neither.—M.G.

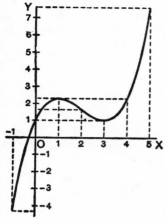

FIG. 28.

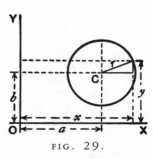

FIG. 29.

Now we know beforehand, by mere inspection of the figure, that when $x = a$, y will be either at its maximum value, $b + r$, or else at its minimum value, $b - r$. But let us not take advantage of this knowledge; let us set about finding what value of x will make y a maximum or a minimum, by the process of differentiating and equating to zero.

$$\frac{dy}{dx} = \frac{1}{2}\frac{1}{\sqrt{r^2 - (x-a)^2}} \times (2a - 2x)$$

which reduces to

$$\frac{dy}{dx} = \frac{a - x}{\sqrt{r^2 - (x-a)^2}}$$

Then the condition for y being maximum or minimum is:

$$\frac{a - x}{\sqrt{r^2 - (x-a)^2}} = 0$$

Since no value whatever of x will make the denominator infinite, the only condition to give zero is

$$x = a$$

Inserting this value in the original equation for the circle, we find

$$y = \sqrt{r^2} + b$$

and as the root of r^2 is either $+r$ or $-r$, we have two resulting values of y,

$$\begin{cases} y = b + r \\ y = b - r \end{cases}$$

The first of these is the maximum, at the top; the second the minimum, at the bottom.

If the curve is such that there is no place that is a maximum or minimum, the process of equating to zero will yield an impossible result. For instance:

Let $\qquad\qquad y = ax^3 + bx + c$

Then $\qquad\qquad \dfrac{dy}{dx} = 3ax^2 + b$

Equating this to zero, we get $3ax^2 + b = 0$, $x^2 = \dfrac{-b}{3a}$, and $x = \sqrt{\dfrac{-b}{3a}}$, which is impossible, supposing a and b to have the same sign.

same sign.

Therefore y has no maximum nor minimum.

A few more worked examples will enable you to thoroughly master this most interesting and useful application of the calculus.[2]

(1) What are the sides of the rectangle of maximum area inscribed in a circle of radius R?

---

2. This is an understatement. Finding extrema (maxima and minima) values of a function is one of the most beautiful and useful aspects of differential calculus. Simply equate a derivative to zero and solve for x! It seems to work like magic!—M.G.

If one side be called x,

$$\text{the other side} = \sqrt{(\text{diagonal})^2 - x^2};$$

and as the diagonal of the rectangle is necessarily a diameter of the circumscribing circle, the other side $=\sqrt{4R^2 - x^2}$.

Then, area of rectangle $S = x\sqrt{4R^2 - x^2}$,

$$\frac{dS}{dx} = x \times \frac{d\left(\sqrt{4R^2 - x^2}\right)}{dx} + \sqrt{4R^2 - x^2} \times \frac{d(x)}{dx}$$

If you have forgotten how to differentiate $\sqrt{4R^2 - x^2}$, here is a hint: write $w = 4R^2 - x^2$ and $y = \sqrt{w}$, and seek $\dfrac{dy}{dw}$ and $\dfrac{dw}{dx}$; fight it out, and only if you can't get on refer to Chapter IX.

You will get

$$\frac{dS}{dx} = x \times -\frac{x}{\sqrt{4R^2 - x^2}} + \sqrt{4R^2 - x^2} = \frac{4R^2 - 2x^2}{\sqrt{4R^2 - x^2}}$$

For maximum or minimum we must have

$$\frac{4R^2 - 2x^2}{\sqrt{4R^2 - x^2}} = 0$$

that is, $4R^2 - 2x^2 = 0$ and $x = R\sqrt{2}$.

The other side $= \sqrt{4R^2 - 2R^2} = R\sqrt{2}$; the two sides are equal; the figure is a square the side of which is equal to the diagonal of the square constructed on the radius. In this case it is, of course, a maximum with which we are dealing.

(2) What is the radius of the opening of a conical vessel the sloping side of which has a length l when the capacity of the vessel is greatest?

If R be the radius and H the corresponding height,

$$H = \sqrt{l^2 - R^2}$$

Volume $V = \pi R^2 \times \dfrac{H}{3} = \pi R^2 \times \dfrac{\sqrt{l^2 - R^2}}{3}$

Proceeding as in the previous problem, we get

$$\frac{dV}{dR} = \pi R^2 \times -\frac{R}{3\sqrt{l^2 - R^2}} + \frac{2\pi R}{3}\sqrt{l^2 - R^2}$$

$$= \frac{2\pi R(l^2 - R^2) - \pi R^3}{3\sqrt{l^2 - R^2}} = 0$$

for maximum or minimum.

Or, $2\pi R(l^2 - R^2) - \pi R^3 = 0$, and $R = l\sqrt{\frac{2}{3}}$, for a maximum, obviously.

(3) Find the maxima and minima of the function

$$y = \frac{x}{4-x} + \frac{4-x}{x}$$

We get $\quad \dfrac{dy}{dx} = \dfrac{(4-x)-(-x)}{(4-x)^2} + \dfrac{-x-(4-x)}{x^2} = 0$

for maximum or minimum; or

$$\frac{4}{(4-x)^2} - \frac{4}{x^2} = 0 \quad \text{and} \quad x = 2$$

There is only one value, hence only one maximum or minimum.

$$\text{For} \quad x = 2 \qquad y = 2$$
$$\text{for} \quad x = 1.5 \quad y = 2.27$$
$$\text{for} \quad x = 2.5 \quad y = 2.27$$

it is therefore a minimum. (It is instructive to plot the graph of the function.)

(4) Find the maxima and minima of the function

$$y = \sqrt{1+x} + \sqrt{1-x}$$

(It will be found instructive to plot the graph.)

Differentiating gives at once (see example No. 1, Chapter IX).

$$\frac{dy}{dx} = \frac{1}{2\sqrt{1+x}} - \frac{1}{2\sqrt{1-x}} = 0$$

for maximum or minimum.

Hence $\sqrt{1+x} = \sqrt{1-x}$ and $x = 0$, the only solution

For $x = 0$, $y = 2$.

For $x = \pm0.5$, $y = 1.932$, so this is a maximum.

(5) Find the maxima and minima of the function

$$y = \frac{x^2 - 5}{2x - 4}$$

We have $\dfrac{dy}{dx} = \dfrac{(2x - 4) \times 2x - (x^2 - 5)2}{(2x - 4)^2} = 0$

for maximum or minimum; or

$$\frac{2x^2 - 8x + 10}{(2x - 4)^2} = 0$$

or $x^2 - 4x + 5 = 0$; which has solutions

$$x = 2 \pm \sqrt{-1}$$

These being imaginary, there is no real value of x for which $\dfrac{dy}{dx} = 0$; hence there is neither maximum nor minimum.

(6) Find the maxima and minima of the function

$$(y - x^2)^2 = x^5$$

This may be written $y = x^2 \pm x^{\frac{5}{2}}$.

$$\frac{dy}{dx} = 2x \pm \tfrac{5}{2}x^{\frac{3}{2}} = 0 \text{ for maximum or minimum;}$$

that is, $x\left(2 \pm \tfrac{5}{2}x^{\frac{1}{2}}\right) = 0$, which is satisfied for $x = 0$, and for $2 \pm \tfrac{5}{2}x^{\frac{1}{2}} = 0$, that is for $x = \frac{16}{25}$. So there are two solutions.

Taking first $x = 0$. If $x = -0.5$, $y = 0.25 \pm \sqrt[2]{-(.5)^5}$, and if $x = +0.5$, $y = 0.25 \pm \sqrt[2]{(.5)^5}$. On one side y is imaginary; that is, there is no value of y that can be represented by a graph; the latter is therefore entirely on the right side of the axis of y (see Fig. 30).

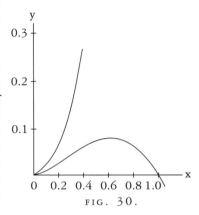

On plotting the graph it will be found that the curve goes to the origin, as if there were a minimum there; but instead of continuing beyond, as it should do for a minimum, it retraces its steps (forming a cusp). There is no minimum, therefore, although the condition for a minimum is satisfied, namely $\frac{dy}{dx} = 0$. It is necessary therefore always to check by taking one value on either side.[3]

Now, if we take $x = \frac{16}{25} = 0.64$. If $x = 0.64$, $y = 0.7373$ and $y = 0.0819$; if $x = 0.6$, y becomes 0.6389 and 0.0811; and if $x = 0.7$, y becomes 0.9000 and 0.0800.

This shows that there are two branches of the curve; the upper one does not pass through a maximum, but the lower one does.

[At this point Thompson introduces a problem that has nothing to do with extrema, but I have let it remain. Note also that problem 10 of the exercises is also out of place for the same reason.—M.G.]

(7) A cylinder whose height is twice the radius of the base is increasing in volume, so that all its parts keep always in the same proportion to each other; that is, at any instant, the cylinder is *similar* to the original cylinder. When the radius of the base is r inches, the surface area is increasing at the rate of 20 square inches per second; at what rate per second is its volume then increasing?

$$\text{Area} = S = 2(\pi r^2) + 2\pi r \times 2r = 6\pi r^2$$

$$\text{Volume} = V = \pi r^2 \times 2r = 2\pi r^3$$

3. Today's terminology broadens the meaning of extrema, allowing local extrema at cusps, corners, and endpoints of intervals in a function's domain.—M.G.

$$\frac{dS}{dt} = 12\pi r \frac{dr}{dt} = 20; \quad \frac{dr}{dt} = \frac{20}{12\pi r}$$

$$\frac{dV}{dt} = 6\pi r^2 \frac{dr}{dt}; \quad \text{and}$$

$$\frac{dV}{dt} = 6\pi r^2 \times \frac{20}{12\pi r} = 10r$$

The volume changes at the rate of $10r$ cubic inches per second.

Make other examples for yourself. There are few subjects which offer such a wealth for interesting examples.

EXERCISES IX

(1) What values of x will make y a maximum and a minimum, if $y = \dfrac{x^2}{x + 1}$?

(2) What value of x will make y a maximum in the equation $y = \dfrac{x}{a^2 + x^2}$?

(3) A line of length p is to be cut up into 4 parts and put together as a rectangle. Show that the area of the rectangle will be a maximum if each of its sides is equal to $\frac{1}{4}p$.

(4) A piece of string 30 inches long has its two ends joined together and is stretched by 3 pegs so as to form a triangle. What is the largest triangular area that can be enclosed by the string?

(5) Plot the curve corresponding to the equation

$$y = \frac{10}{x} + \frac{10}{8 - x}$$

also find $\dfrac{dy}{dx}$, and deduce the value of x that will make y a minimum; and find that minimum value of y.

(6) If $y = x^5 - 5x$, find what values of x will make y a maximum or a minimum.

(7) What is the smallest square that can be placed in a given square so each corner of the small square touches a side of the larger square?

(8) Inscribe in a given cone, the height of which is equal to the radius of the base, a cylinder (*a*) whose volume is a maximum; (*b*) whose lateral area is a maximum; (*c*) whose total area is a maximum.

(9) Inscribe in a sphere, a cylinder (*a*) whose volume is a maximum; (*b*) whose lateral area is a maximum; (*c*) whose total area is a maximum.

(10) A spherical balloon is increasing in volume. If, when its radius is *r* feet, its volume is increasing at the rate of 4 cubic feet per second, at what rate is its surface then increasing?

(11) Inscribe in a given sphere a cone whose volume is a maximum.

Chapter XII

CURVATURE OF CURVES

Returning to the process of successive differentiation, it may be asked: Why does anybody want to differentiate twice over? We know that when the variable quantities are space and time, by differentiating twice over we get the acceleration of a moving body, and that in the geometrical interpretation, as applied to curves, $\frac{dy}{dx}$ means the *slope* of the curve. But what can $\frac{d^2y}{dx^2}$ mean in this case? Clearly it means the rate (per unit of length x) at which the slope is changing—in brief, it is *an indication of the manner in which the slope of the portion of curve considered varies,* that is, whether the slope of the curve increases or decreases when x increases, or, in other words, whether the curve curves up or down towards the right.

Suppose a slope constant, as in Fig. 31.

Here, $\frac{dy}{dx}$ is of constant value.

Suppose, however, a case in which, like Fig. 32, the slope itself

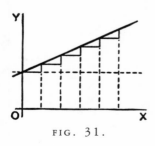

FIG. 31.

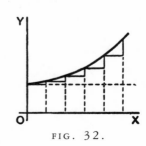

FIG. 32.

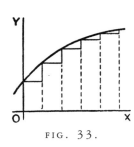

FIG. 33.

is getting greater upwards; then $\dfrac{d\left(\dfrac{dy}{dx}\right)}{dx}$,

that is, $\dfrac{d^2y}{dx^2}$, will be *positive.*

If the slope is becoming less as you go to the right (as in Fig. 14), or as in Fig. 33, then, even though the curve may be going upward, since the change is such as to diminish its slope, its $\dfrac{d^2y}{dx^2}$ will be *negative.*

It is now time to initiate you into another secret—how to tell whether the result that you get by "equating to zero" is a maximum or a minimum. The trick is this: After you have differentiated (so as to get the expression which you equate to zero), you then differentiate a second time and look whether the result of the second differentiation is *positive* or *negative.* If $\dfrac{d^2y}{dx^2}$ comes out *positive,* then you know that the value of y which you got was a *minimum*; but if $\dfrac{d^2y}{dx^2}$ comes out *negative,* then the value of y which you got must be a *maximum.* That's the rule.

The reason of it ought to be quite evident. Think of any curve that has a minimum point in it, like Fig. 15, or like Fig. 34, where the point of minimum y is marked M, and the curve is *concave* upward. To the left of M the slope is downward, that is, negative, and is getting less negative. To the right of M the slope has become upward, and is getting more and more upward. Clearly the change of slope as the curve passes through M is such that $\dfrac{d^2y}{dx^2}$ is *positive,* for its operation, as x increases toward the right, is to convert a downward slope into an upward one.

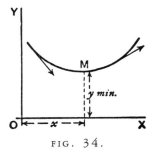

FIG. 34.

Similarly, consider any curve that has a maximum point in it, like Fig. 16, or like Fig. 35, where the curve is *concave upward,* and the maximum point is marked M. In this case, as the curve passes through M from left to right, its upward slope is converted into a downward or negative slope, so that in this case the "slope of the slope" $\dfrac{d^2y}{dx^2}$ is *negative.*

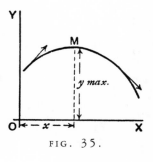

FIG. 35.

Go back now to the examples of the last chapter and verify in this way the conclusions arrived at as to whether in any particular case there is a maximum or a minimum. You will find below a few worked-out examples.

(1) Find the maximum or minimum of

(*a*) $y = 4x^2 - 9x - 6$;   (*b*) $y = 6 + 9x - 4x^2$

and ascertain if it be a maximum or a minimum in each case.

(*a*) $\dfrac{dy}{dx} = 8x - 9 = 0$; $x = 1\tfrac{1}{8}$; and $y = -11.0625$

$\dfrac{d^2y}{dx^2} = 8$; it is +; hence it is a minimum.

(*b*) $\dfrac{dy}{dx} = 9 - 8x = 0$; $x = 1\tfrac{1}{8}$; and $y = +11.0625$

$\dfrac{d^2y}{dx^2} = -8$; it is −; hence it is a maximum.

(2) Find the maxima and minima of the function

$$y = x^3 - 3x + 16$$

$$\dfrac{dy}{dx} = 3x^2 - 3 = 0; \quad x^2 = 1; \text{ and } x = \pm 1$$

$$\frac{d^2y}{dx^2} = 6x; \text{ for } x = 1; \text{ it is } +,$$

hence $x = 1$ corresponds to a minimum $y = 14$. For $x = -1$ it is $-$; hence $x = -1$ corresponds to a maximum $y = +18$.

(3) Find the maxima and minima of $y = \dfrac{x - 1}{x^2 + 2}$.

$$\frac{dy}{dx} = \frac{(x^2 + 2) \times 1 - (x - 1) \times 2x}{(x^2 + 2)^2} = \frac{2x - x^2 + 2}{(x^2 + 2)^2} = 0$$

or $x^2 - 2x - 2 = 0$, whose solutions are $x = +2.73$ and $x = -0.73$.

$$\frac{d^2y}{dx^2} = -\frac{(x^2 + 2)^2(2 - 2x) - (2x - x^2 + 2)(4x^3 + 8x)}{(x^2 + 2)^4}$$

$$= \frac{2x^5 - 6x^4 - 8x^3 - 8x^2 - 24x + 8}{(x^2 + 2)^4}$$

The denominator is always positive, so it is sufficient to ascertain the sign of the numerator.

If we put $x = 2.73$, the numerator is negative; the maximum, $y = 0.183$.

If we put $x = -0.73$, the numerator is positive; the minimum, $y = -0.683$.

(4) The expense C of handling the products of a certain factory varies with the weekly output P according to the relation $C = aP + \dfrac{b}{c + P} + d$, where a, b, c, d are positive constants. For what output will the expense be least?

$$\frac{dC}{dP} = a - \frac{b}{(c + P)^2} = 0 \text{ for maximum or minimum; hence}$$

$$a = \frac{b}{(c + P)^2} \text{ and } P = \pm\sqrt{\frac{b}{a}} - c$$

As the output cannot be negative, $P = +\sqrt{\dfrac{b}{a}} - c$.

Now $$\frac{d^2C}{dP^2} = +\frac{b(2c + 2P)}{(c + P)^4}$$

which is positive for all the values of P; hence $P = +\sqrt{\dfrac{b}{a}} - c$ corresponds to a minimum.

(5) The total cost per hour C of lighting a building with N lamps of a certain kind is

$$C = N\left(\frac{C_l}{t} + \frac{EPC_e}{1000}\right)$$

where E is the commercial efficiency (watts per candle),

P is the candle power of each lamp,

t is the average life of each lamp in hours,

$C_l = $ cost of renewal in cents per hour of use,

$C_e = $ cost of energy per 1000 watts per hour.

Moreover, the relation connecting the average life of a lamp with the commercial efficiency at which it is run is approximately $t = mE^n$, where m and n are constants depending on the kind of lamp.

Find the commercial efficiency for which the total cost of lighting will be least.

We have $$C = N\left(\frac{C_l}{m}E^{-n} + \frac{PC_e}{1000}E\right)$$

$$\frac{dC}{dE} = N\left(\frac{PC_e}{1000} - \frac{nC_l}{m}E^{-(n+1)}\right) = 0$$

for maximum or minimum.

$$E^{n+1} = \frac{1000 \times nC_l}{mPC_e} \quad \text{and} \quad E = \sqrt[n+1]{\frac{1000 \times nC_l}{mPC_e}}$$

This is clearly for minimum, since

$$\frac{d^2C}{dE^2} = N\left[(n+1)\frac{nC_l}{m}E^{-(n+2)}\right]$$

which is positive for a positive value of E.

For a particular type of 16 candle-power lamps, $C_l = 17$ cents, $C_e = 5$ cents; and it was found that $m = 10$ and $n = 3.6$.

$$E = \sqrt[4.6]{\frac{1000 \times 3.6 \times 17}{10 \times 16 \times 5}} = 2.6 \text{ watts per candle power.}$$

EXERCISES X

You are advised to plot the graph of any numerical example.

(1) Find the maxima and minima of

$$y = x^3 + x^2 - 10x + 8$$

(2) Given $y = \frac{b}{a}x - cx^2$, find expressions for $\frac{dy}{dx}$, and for $\frac{d^2y}{dx^2}$;

also find the value of x which makes y a maximum or a minimum, and show whether it is maximum or minimum. Assume $c > 0$.

(3) Find how many maxima and how many minima there are in the curve, the equation to which is

$$y = 1 - \frac{x^2}{2} + \frac{x^4}{24}$$

and how many in that of which the equation is

$$y = 1 - \frac{x^2}{2} + \frac{x^4}{24} - \frac{x^6}{720}$$

(4) Find the maxima and minima of

$$y = 2x + 1 + \frac{5}{x^2}$$

(5) Find the maxima and minima of

$$y = \frac{3}{x^2 + x + 1}$$

(6) Find the maxima and minima of

$$y = \frac{5x}{2 + x^2}$$

(7) Find the maxima and minima of

$$y = \frac{3x}{x^2 - 3} + \frac{x}{2} + 5$$

(8) Divide a number N into two parts in such a way that three times the square of one part plus twice the square of the other part shall be a minimum.

(9) The efficiency u of an electric generator at different values of output x is expressed by the general equation:

$$u = \frac{x}{a + bx + cx^2}$$

where a is a constant depending chiefly on the energy losses in the iron and c a constant depending chiefly on the resistance of the copper parts. Find an expression for that value of the output at which the efficiency will be a maximum.

(10) Suppose it to be known that consumption of coal by a certain steamer may be represented by the formula

$$y = 0.3 + 0.001v^3$$

where y is the number of tons of coal burned per hour and v is the speed expressed in nautical miles per hour. The cost of wages, interest on capital, and depreciation of that ship are together equal, per hour, to the cost of 1 ton of coal. What speed will make the total cost of a voyage of 1000 nautical miles a minimum? And, if coal costs 10 dollars per ton, what will that minimum cost of the voyage amount to?

(11) Find the maxima and minima of

$$y = \pm\frac{x}{6}\sqrt{x(10 - x)}$$

(12) Find the maxima and minima of

$$y = 4x^3 - x^2 - 2x + 1$$

Chapter XIII

PARTIAL FRACTIONS AND INVERSE FUNCTIONS

We have seen that when we differentiate a fraction we have to perform a rather complicated operation; and, if the fraction is not itself a simple one, the result is bound to be a complicated expression. If we could split the fraction into two or more simpler fractions such that their sum is equivalent to the original fraction, we could then proceed by differentiating each of these simpler expressions. And the result of differentiating would be the sum of two (or more) derivatives, each one of which is relatively simple; while the final expression, though of course it will be the same as that which could be obtained without resorting to this dodge, is thus obtained with much less effort and appears in a simplified form.

Let us see how to reach this result. Try first the job of adding two fractions together to form a resultant fraction. Take, for example, the two fractions $\dfrac{1}{x+1}$ and $\dfrac{2}{x-1}$. Every schoolboy can add these together and find their sum to be $\dfrac{3x+1}{x^2-1}$. And in the same way he can add together three or more fractions. Now this process can certainly be reversed: that is to say that, if this last expression were given, it is certain that it can somehow be split back again into its original components or partial fractions. Only we do not know in every case that may be presented to us *how* we can so split it. In order to find this out we shall consider a simple case at first. But it is important to bear in mind that all which follows applies only to what are called "proper" algebraic fractions, meaning fractions like the above, which have the numerator of *a lesser degree* than the denominator; that is, those in which the highest exponent of x is less in the numerator than in the denominator. If we have to deal with such an expression as $\dfrac{x^2 + 2}{x^2 - 1}$, we can simplify it by division, since it is equivalent to $1 + \dfrac{3}{x^2 - 1}$; and $\dfrac{3}{x^2 - 1}$ is a proper algebraic fraction to which the operation of splitting into partial fractions can be applied, as explained hereafter.

Case I. If we perform many additions of two or more fractions the denominators of which contain only terms in x, and no terms in x^2, x^3, or any other powers of x, we *always* find that *the denominator of the final resulting fraction is the product of the denominators* of the fractions which were added to form the result. It follows that by factorizing the denominator of this final fraction, we can find every one of the denominators of the partial fractions of which we are in search.

Suppose we wish to go back from $\dfrac{3x + 1}{x^2 - 1}$ to the components which we know are $\dfrac{1}{x + 1}$ and $\dfrac{2}{x - 1}$. If we did not know what those components were we can still prepare the way by writing:

$$\frac{3x + 1}{x^2 - 1} = \frac{3x + 1}{(x + 1)(x - 1)} = \frac{}{x + 1} + \frac{}{x - 1}$$

leaving blank the places for the numerators until we know what to put there. We always may assume the sign between the partial fractions to be *plus,* since, if it be *minus,* we shall simply find the corresponding numerator to be negative. Now, since the partial fractions are *proper* fractions, the numerators are mere numbers without x at all, and we can call them $A, B, C \ldots$ as we please. So, in this case, we have:

$$\frac{3x + 1}{x^2 - 1} = \frac{A}{x + 1} + \frac{B}{x - 1}$$

If, now, we perform the addition of these two partial fractions, we get $\dfrac{A(x-1)+B(x+1)}{(x+1)(x-1)}$; and this must be equal to $\dfrac{3x+1}{(x+1)(x-1)}$. And, as the denominators in these two expressions are the same, the numerators must be equal, giving us:

$$3x+1 = A(x-1) + B(x+1)$$

Now, this is an equation with two unknown quantities, and it would seem that we need another equation before we can solve them and find A and B. But there is another way out of this difficulty. The equation must be true for all values of x; therefore it must be true for such values of x as will cause $x-1$ and $x+1$ to become zero, that is for $x=1$ and for $x=-1$ respectively. If we make $x=1$, we get $4 = (A \times 0) + (B \times 2)$, so that $B=2$; and if we make $x=-1$, we get

$$-2 = (A \times -2) + (B \times 0)$$

so that $A=1$. Replacing the A and B of the partial fractions by these new values, we find them to become $\dfrac{1}{x+1}$ and $\dfrac{2}{x-1}$; and the thing is done.

As a further example, let us take the fraction

$$\frac{4x^2 + 2x - 14}{x^3 + 3x^2 - x - 3}$$

The denominator becomes zero when x is given the value 1; hence $x-1$ is a factor of it, and obviously then the other factor will be $x^2 + 4x + 3$; and this can again be decomposed into $(x+1)(x+3)$. So we may write the fraction thus:

$$\frac{4x^2 + 2x - 14}{x^3 + 3x^2 - x - 3} = \frac{A}{x+1} + \frac{B}{x-1} + \frac{C}{x+3}$$

making three partial factors.

Proceeding as before, we find

$$4x^2 + 2x - 14 = A(x-1)(x+3) + B(x+1)(x+3) + C(x+1)(x-1)$$

Now, if we make $x = 1$, we get:

$$-8 = (A \times 0) + B(2 \times 4) + (C \times 0); \text{ that is, } B = -1$$

If $x = -1$, we get

$$-12 = A(-2 \times 2) + (B \times 0) + (C \times 0); \text{ whence } A = 3.$$

If $x = -3$, we get:

$$16 = (A \times 0) + (B \times 0) + C(-2 \times -4); \text{ whence } C = 2.$$

So then the partial fractions are:

$$\frac{3}{x+1} - \frac{1}{x-1} + \frac{2}{x+3}$$

which is far easier to differentiate with respect to x than the complicated expression from which it is derived.

Case II. If some of the factors of the denominator contain terms in x^2, and are not conveniently put into factors, then the corresponding numerator may contain a term in x, as well as a simple number, and hence it becomes necessary to represent this unknown numerator not by the symbol A but by $Ax + B$; the rest of the calculation being made as before.

Try, for instance: $\dfrac{-x^2 - 3}{(x^2 + 1)(x + 1)}$

$$\frac{-x^2 - 3}{(x^2 + 1)(x + 1)} = \frac{Ax + B}{x^2 + 1} + \frac{C}{x + 1}$$

$$-x^2 - 3 = (Ax + B)(x + 1) + C(x^2 + 1)$$

Putting $x = -1$, we get $-4 = C \times 2$; and $C = -2$

hence $\qquad -x^2 - 3 = (Ax + B)(x + 1) - 2x^2 - 2$

and $\qquad x^2 - 1 = Ax(x + 1) + B(x + 1)$

Putting $x = 0$, we get $-1 = B$;

hence $x^2 - 1 = Ax(x + 1) - x - 1;$ or $x^2 + x = Ax(x + 1)$

and $\qquad\qquad x + 1 = A(x + 1)$

so that $A = 1$, and the partial fractions are:

$$\frac{x-1}{x^2+1} - \frac{2}{x+1}$$

Take as another example the fraction

$$\frac{x^3 - 2}{(x^2+1)(x^2+2)}$$

We get
$$\frac{x^3-2}{(x^2+1)(x^2+2)} = \frac{Ax+B}{x^2+1} + \frac{Cx+D}{x^2+2}$$

$$= \frac{(Ax+B)(x^2+2) + (Cx+D)(x^2+1)}{(x^2+1)(x^2+2)}$$

In this case the determination of A, B, C, D is not so easy. It will be simpler to proceed as follows: Since the given fraction and the fraction found by adding the partial fractions are equal, and have *identical* denominators, the numerators must also be identically the same. In such a case, and for such algebraical expressions as those with which we are dealing here, *the coefficients of the same powers of x are equal and of same sign.*

Hence, since

$$x^3 - 2 = (Ax+B)(x^2+2) + (Cx+D)(x^2+1)$$

$$= (A+C)x^3 + (B+D)x^2 + (2A+C)x + 2B + D$$

we have $1 = A + C$; $0 = B + D$ (the coefficient of x^2 in the left expression being zero); $0 = 2A + C$; and $-2 = 2B + D$. Here are four equations, from which we readily obtain $A = -1$; $B = -2$; $C = 2$; $D = 2$; so that the partial fractions are $\dfrac{2(x+1)}{x^2+2} - \dfrac{x+2}{x^2+1}$. This method can always be used; but the method shown first will be found the quickest in the case of factors in x only.

Case III. When among the factors of the denominator there are some which are raised to some power, one must allow for the possible existence of partial fractions having for denominator the several powers of that factor up to the highest. For instance, in splitting the fraction $\dfrac{3x^2 - 2x + 1}{(x + 1)^2(x - 2)}$ we must allow for the possible existence of a denominator $x + 1$ as well as $(x + 1)^2$ and $x - 2$.

It may be thought, however, that, since the numerator of the fraction the denominator of which is $(x + 1)^2$ may contain terms in x, we must allow for this in writing $Ax + B$ for its numerator, so that

$$\frac{3x^2 - 2x + 1}{(x + 1)^2(x - 2)} = \frac{Ax + B}{(x + 1)^2} + \frac{C}{x + 1} + \frac{D}{x - 2}$$

If, however, we try to find A, B, C and D in this case, we fail, because we get four unknowns; and we have only three relations connecting them, yet

$$\frac{3x^2 - 2x + 1}{(x + 1)^2(x - 2)} = \frac{x - 1}{(x + 1)^2} + \frac{1}{x + 1} + \frac{1}{x - 2}$$

But if we write

$$\frac{3x^2 - 2x + 1}{(x + 1)^2(x - 2)} = \frac{A}{(x + 1)^2} + \frac{B}{x + 1} + \frac{C}{x - 2}$$

we get $\quad 3x^2 - 2x + 1 = A(x - 2) + B(x + 1)(x - 2) + C(x + 1)^2$.

For $\quad x = -1; \quad 6 = -3A, \quad$ or $A = -2$

For $\quad x = 2; \quad 9 = 9C, \quad$ or $C = 1$

For $\quad x = 0; \quad 1 = -2A - 2B + C$

Putting in the values of A and C:

$$1 = 4 - 2B + 1, \text{ from which } B = 2$$

Hence the partial fractions are:

$$\frac{2}{x + 1} - \frac{2}{(x + 1)^2} + \frac{1}{x - 2}$$

instead of $\dfrac{1}{x+1} + \dfrac{x-1}{(x+1)^2} + \dfrac{1}{x-2}$ stated above as being the fractions from which $\dfrac{3x^2 - 2x + 1}{(x+1)^2(x-2)}$ was obtained. The mystery is cleared if we observe that $\dfrac{x-1}{(x+1)^2}$ can itself be split into the two fractions $\dfrac{1}{x+1} - \dfrac{2}{(x+1)^2}$, so that the three fractions given are really equivalent to

$$\frac{1}{x+1} + \frac{1}{x+1} - \frac{2}{(x+1)^2} + \frac{1}{x-2} = \frac{2}{x+1} - \frac{2}{(x+1)^2} + \frac{1}{x-2}$$

which are the partial fractions obtained.

We see that it is sufficient to allow for one numerical term in each numerator, and that we always get the ultimate partial fractions.

When there is a power of a factor of x^2 in the denominator, however, the corresponding numerators must be of the form $Ax + B$; for example,

$$\frac{3x - 1}{(2x^2 - 1)^2(x + 1)} = \frac{Ax + B}{(2x^2 - 1)^2} + \frac{Cx + D}{2x^2 - 1} + \frac{E}{x + 1}$$

which gives

$$3x - 1 = (Ax + B)(x + 1) + (Cx + D)(x + 1)(2x^2 - 1) + E(2x^2 - 1)^2.$$

For $x = -1$, this gives $E = -4$. Replacing, transposing, collecting like terms, and dividing by $x + 1$, we get

$$16x^3 - 16x^2 + 3 = 2Cx^3 + 2Dx^2 + x(A - C) + (B - D).$$

Hence $2C = 16$ and $C = 8$; $2D = -16$ and $D = -8$; $A - C = 0$ or $A - 8 = 0$ and $A = 8$; and finally, $B - D = 3$ or $B = -5$. So that we obtain as the partial fractions:

$$\frac{8x - 5}{(2x^2 - 1)^2} + \frac{8(x - 1)}{2x^2 - 1} - \frac{4}{x + 1}$$

It is useful to check the results obtained. The simplest way is to replace x by a single value, say $+1$, both in the given expression and in the partial fractions obtained.

Whenever the denominator contains but a power of a single factor, a very quick method is as follows:

Taking, for example, $\dfrac{4x + 1}{(x + 1)^3}$, let $x + 1 = z$; then $x = z - 1$.

Replacing, we get

$$\frac{4(z - 1) + 1}{z^3} = \frac{4z - 3}{z^3} = \frac{4}{z^2} - \frac{3}{z^3}$$

The partial fractions are, therefore,

$$\frac{4}{(x + 1)^2} - \frac{3}{(x + 1)^3}$$

Applying this to differentiation, let it be required to differentiate $y = \dfrac{5 - 4x}{6x^2 + 7x - 3}$; we have

$$\frac{dy}{dx} = -\frac{(6x^2 + 7x - 3) \times 4 + (5 - 4x)(12x + 7)}{(6x^2 + 7x - 3)^2}$$

$$= \frac{24x^2 - 60x - 23}{(6x^2 + 7x - 3)^2}$$

If we split the given expression into

$$\frac{1}{3x - 1} - \frac{2}{2x + 3}$$

we get, however,

$$\frac{dy}{dx} = -\frac{3}{(3x - 1)^2} + \frac{4}{(2x + 3)^2}$$

which is really the same result as above split into partial fractions. But the splitting, if done after differentiating, is more complicated, as will easily be seen. When we shall deal with the *integration* of such expressions, we shall find the splitting into partial fractions a precious auxiliary.

EXERCISES XI

Split into partial fractions:

(1) $\dfrac{3x + 5}{(x - 3)(x + 4)}$

(2) $\dfrac{3x - 4}{(x - 1)(x - 2)}$

(3) $\dfrac{3x + 5}{x^2 + x - 12}$

(4) $\dfrac{x + 1}{x^2 - 7x + 12}$

(5) $\dfrac{x - 8}{(2x + 3)(3x - 2)}$

(6) $\dfrac{x^2 - 13x + 26}{(x - 2)(x - 3)(x - 4)}$

(7) $\dfrac{x^2 - 3x + 1}{(x - 1)(x + 2)(x - 3)}$

(8) $\dfrac{5x^2 + 7x + 1}{(2x + 1)(3x - 2)(3x + 1)}$

(9) $\dfrac{x^2}{x^3 - 1}$

(10) $\dfrac{x^4 + 1}{x^3 + 1}$

(11) $\dfrac{5x^2 + 6x + 4}{(x + 1)(x^2 + x + 1)}$

(12) $\dfrac{x}{(x - 1)(x - 2)^2}$

(13) $\dfrac{x}{(x^2 - 1)(x + 1)}$

(14) $\dfrac{x + 3}{(x + 2)^2(x - 1)}$

(15) $\dfrac{3x^2 + 2x + 1}{(x + 2)(x^2 + x + 1)^2}$

(16) $\dfrac{5x^2 + 8x - 12}{(x + 4)^3}$

(17) $\dfrac{7x^2 + 9x - 1}{(3x - 2)^4}$

(18) $\dfrac{x^2}{(x^3 - 8)(x - 2)}$

Derivative of an Inverse Function

Consider the function $y = 3x$; it can be expressed in the form $x = \dfrac{y}{3}$; this latter form is called the *inverse function* to the one originally given.

If $y = 3x$, $\dfrac{dy}{dx} = 3$; if $x = \dfrac{y}{3}$, $\dfrac{dx}{dy} = \dfrac{1}{3}$, and we see that

$$\frac{dy}{dx} = \frac{1}{\dfrac{dx}{dy}} \quad \text{or} \quad \frac{dy}{dx} \times \frac{dx}{dy} = 1$$

Consider $y = 4x^2$, $\dfrac{dy}{dx} = 8x$; the inverse function is

$$x = \frac{y^{\frac{1}{2}}}{2}, \quad \text{and} \quad \frac{dx}{dy} = \frac{1}{4\sqrt{y}} = \frac{1}{4 \times 2x} = \frac{1}{8x}$$

Here again $$\frac{dy}{dx} \times \frac{dx}{dy} = 1$$

It can be shown that for all functions which can be put into the inverse form, one can always write

$$\frac{dy}{dx} \times \frac{dx}{dy} = 1 \quad \text{or} \quad \frac{dy}{dx} = \frac{1}{\dfrac{dx}{dy}}$$

It follows that, being given a function, if it be easier to differentiate the inverse function, this may be done, and the reciprocal of the derivative of the inverse function gives the derivative of the given function itself.

As an example, suppose that we wish to differentiate $y = \sqrt{\dfrac{3}{x} - 1}$. We have seen one way of doing this, by writing $u = \dfrac{3}{x} - 1$, and finding $\dfrac{dy}{du}$ and $\dfrac{du}{dx}$. This gives

$$\frac{dy}{dx} = -\frac{3}{2x^2 \sqrt{\dfrac{3}{x} - 1}}$$

If we had forgotten how to proceed by this method, or wished to check our result by some other way of obtaining the derivative, or for any other reason we could not use the ordinary method, we could proceed as follows: The inverse function is $x = \dfrac{3}{1 + y^2}$.

$$\frac{dx}{dy} = -\frac{3 \times 2y}{(1 + y^2)^2} = -\frac{6y}{(1 + y^2)^2}$$

hence

$$\frac{dy}{dx} = \frac{1}{\dfrac{dx}{dy}} = -\frac{(1 + y^2)^2}{6y} = -\frac{\left(1 + \dfrac{3}{x} - 1\right)^2}{6 \times \sqrt{\dfrac{3}{x} - 1}} = -\frac{3}{2x^2 \sqrt{\dfrac{3}{x} - 1}}$$

Let us take, as another example, $y = \dfrac{1}{\sqrt[3]{\theta + 5}}$

The inverse function $\theta = \dfrac{1}{y^3} - 5$ or $\theta = y^{-3} - 5$, and

$$\frac{d\theta}{dy} = -3y^{-4} = -3\sqrt[3]{(\theta + 5)^4}$$

It follows that $\dfrac{dy}{d\theta} = -\dfrac{1}{3\sqrt[3]{(\theta + 5)^4}}$, as might have been found otherwise.

We shall find this dodge most useful later on; meanwhile you are advised to become familiar with it by verifying by its means the results obtained in Exercises I (Chapter IV), Nos. 5, 6, 7; Examples (Chapter IX), Nos. 1, 2, 4; and Exercises VI (Chapter IX), Nos. 1, 2, 3 and 4.

You will surely realize from this chapter and the preceding, that in many respects the calculus is an *art* rather than a *science:* an art only to be acquired, as all other arts are, by practice. Hence you should work many examples, and set yourself other examples, to see if you can work them out, until the various artifices become familiar by use.

Chapter XIV

ON TRUE COMPOUND
INTEREST AND THE LAW OF
ORGANIC GROWTH

Let there be a quantity growing in such a way that the increment of its growth, during a given time, shall always be proportional to its own magnitude. This resembles the process of reckoning interest on money at some fixed rate; for the bigger the capital, the bigger the amount of interest on it in a given time.

Now we must distinguish clearly between two cases, in our calculation, according as the calculation is made by what the arithmetic books call "simple interest", or by what they call "compound interest". For in the former case the capital remains fixed, while in the latter the interest is added to the capital, which therefore increases by successive additions.

(1) *At simple interest.* Consider a concrete case. Let the capital at start be $100, and let the rate of interest be 10 percent per annum. Then the increment to the owner of the capital will be $10 every year. Let him go on drawing his interest every year, and hoard it by putting it by in a stocking or locking it up in his safe. Then, if he goes on for 10 years, by the end of that time he will have received 10 increments of $10 each, or $100, making, with the original $100, a total of $200 in all. His property will have doubled itself in 10 years. If the rate of interest had been 5 percent, he would have had to hoard for 20 years to double his property. If it had been only 2 percent, he would have had to hoard for 50 years. It is easy to see that if the value of the yearly interest is

$\frac{1}{n}$ of the capital, he must go on hoarding for n years in order to

double his property.

Or, if y be the original capital, and the yearly interest is $\frac{y}{n}$, then, at the end of n years, his property will be

$$y + n\frac{y}{n} = 2y$$

(2) *At compound interest.* As before, let the owner begin with a capital of \$100, earning interest at the rate of 10 percent per annum; but, instead of hoarding the interest, let it be added to the capital each year, so that the capital grows year by year. Then, at the end of one year, the capital will have grown to \$110; and in the second year (still at 10%) this will earn \$11 interest. He will start the third year with \$121, and the interest on that will be \$12.10; so that he starts the fourth year with \$133.10, and so on. It is easy to work it out, and find that at the end of the ten years the total capital will have grown to more than \$259. In fact, we see that at the end of each year, each dollar will have earned $\frac{1}{10}$ of a dollar, and therefore, if this is always added on, each year multiplies the capital by $\frac{11}{10}$; and if continued for ten years (which will multiply by this factor ten times over) the original capital will be multiplied by 2.59374. Let us put this into symbols. Put y_0 for the original capital; $\frac{1}{n}$ for the fraction added on at each of the n operations; and y_n for the value of the capital at the end of the nth operation. Then

$$y_n = y_0 \left(1 + \frac{1}{n} \right)^n$$

But this mode of reckoning compound interest once a year is really not quite fair; for even during the first year the \$100 ought to have been growing. At the end of half a year it ought to have been at least \$105, and it certainly would have been fairer had the interest for the second half of the year been calculated on \$105. This would be equivalent to calling it 5% per half-year; with 20 operations, therefore, at each of which the capital is multiplied by $\frac{21}{20}$. If reckoned this way, by the end of ten years the capital would have grown to more than \$265; for

$$\left(1 + \tfrac{1}{20} \right)^{20} = 2.653$$

But, even so, the process is still not quite fair; for, by the end of the first month, there will be some interest earned; and a half-yearly reckoning assumes that the capital remains stationary for six months at a time. Suppose we divided the year into 10 parts, and reckon a one percent interest for each tenth of the year. We now have 100 operations lasting over the ten years; or

$$y_n = \$100\left(1 + \tfrac{1}{100}\right)^{100}$$

which works out to $270.48.

Even this is not final. Let the ten years be divided into 1000 periods, each of $\tfrac{1}{100}$ of a year; the interest being $\tfrac{1}{10}$ percent for each such period; then

$$y_n = \$100\left(1 + \tfrac{1}{1000}\right)^{1000}$$

which works out to a little more than $271.69.

Go even more minutely, and divide the ten years into 10,000 parts, each $\tfrac{1}{1000}$ of a year, with interest at $\tfrac{1}{100}$ of 1 percent. Then

$$y_n = \$100\left(1 + \tfrac{1}{10,000}\right)^{10,000}$$

which amounts to about $271.81.

Finally, it will be seen that what we are trying to find is in reality the ultimate value of the expression $\left(1 + \dfrac{1}{n}\right)^n$, which, as we see, is greater than 2; and which, as we take n larger and larger, grows closer and closer to a particular limiting value. However big you make n, the value of this expression grows nearer and nearer to the limit

$$2.71828 \ldots,$$

a number *never to be forgotten.*[1]

---

1. This number is transcendental—an irrational number that is not the root of any polynomial equation with integer coefficients. It was named e by Euler (1707-1783), a famous Swiss mathematician. He was the first to prove that e is the limit of $(1 + 1/x)^x$ as x approaches infinity.

Like all irrationals, the decimal expansion of e (2.71828 18284 59045 . . .) never repeats. (The curious repetition of 1828 is entirely co-

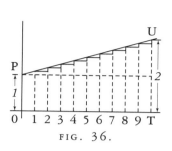

FIG. 36.

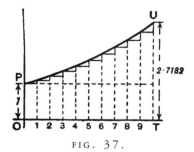

FIG. 37.

Let us take geometrical illustrations of these things. In Fig. 36, *OP* stands for the original value. *OT* is the whole time during which the value is growing. It is divided into 10 periods, in each of which there is an equal step up. Here $\frac{dy}{dx}$ is a constant; and if each step up is $\frac{1}{10}$ of the original *OP*, then, by 10 such steps, the height is doubled. If we had taken 20 steps, each of half the height shown, at the end the height would still be just doubled. Or n such steps, each of $\frac{1}{n}$ of the original height *OP*, would suffice to double the height. This is the case of simple interest. Here is 1 growing till it becomes 2.

In Fig. 37, we have the corresponding illustration of the geometrical progression. Each of the successive ordinates is to be $1 + \frac{1}{n}$, that is, $\frac{n+1}{n}$ times as high as its predecessor. The steps up are not equal, because each step up is now $\frac{1}{n}$ of the ordinate *at that part* of the curve. If we had literally 10 steps, with

incidental.) Its best fractional approximation with integers less than 1,000 is 878/323 = 2.71826. . . .

The number *e* is as ubiquitous as pi, turning up everywhere, and especially in probability theory. If you randomly select real numbers between 0 and 1, and continue until their sum exceeds 1, the expected number of choices is *e*. See the chapter on *e* in my *Unexpected Hanging and Other Mathematical Diversions* (1969).—M.G.

$\left(1 + \frac{1}{10}\right)$ for the multiplying factor, the final total would be $\left(1 + \frac{1}{10}\right)^{10}$ or 2.594 times the original 1. But if only we take n sufficiently large (and the corresponding $\frac{1}{n}$ sufficiently small), then the final value $\left(1 + \frac{1}{n}\right)^{n}$ to which unity will grow will be 2.71828. . . .

e. To this mysterious number 2.7182818. . . , the mathematicians have assigned the English letter *e.* All schoolboys know that the Greek letter π (called *pi*) stands for 3.141592. . . , but how many of them know that *e* stands for 2.71828. . . . Yet it is an even more important number than π!

What, then, is *e?*

Suppose we were to let 1 grow at simple interest till it became 2; then, if at the same nominal rate of interest, and for the same time, we were to let 1 grow at true compound interest, instead of simple, it would grow to the value *e.*

This process of growing proportionately, at every instant, to the magnitude at that instant, some people call an *exponential rate* of growing. Unit exponential rate of growth is that rate which in unit time will cause 1 to grow to 2.718281. . . . It might also be called the *organic rate* of growing: because it is characteristic of organic growth (in certain circumstances) that the increment of the organism in a given time is proportional to the magnitude of the organism itself.

If we take 100 percent as the unit of rate, and any fixed period as the unit of time, then the result of letting 1 grow *arithmetically* at unit rate, for unit time, will be 2, while the result of letting 1 grow *exponentially* at unit rate, for the same time, will be 2.71828. . . .

A Little More About e.   We have seen that we require to know what value is reached by the expression $\left(1 + \frac{1}{n}\right)^{n}$, when n becomes infinitely great. Arithmetically, here are tabulated a lot of values obtained by assuming $n = 2$; $n = 5$; $n = 10$; and so on, up to $n = 10,000$.

$$\left(1 + \tfrac{1}{2}\right)^2 \qquad = 2.25$$

$$\left(1 + \tfrac{1}{5}\right)^5 \qquad = 2.489$$

$$\left(1 + \tfrac{1}{10}\right)^{10} \qquad = 2.594$$

$$\left(1 + \tfrac{1}{20}\right)^{20} \qquad = 2.653$$

$$\left(1 + \tfrac{1}{100}\right)^{100} \qquad = 2.705$$

$$\left(1 + \tfrac{1}{1000}\right)^{1000} \qquad = 2.7169$$

$$\left(1 + \tfrac{1}{10,000}\right)^{10,000} = 2.7181$$

It is, however, worth while to find another way of calculating this immensely important figure.

Accordingly, we will avail ourselves of the binomial theorem, and expand the expression $\left(1 + \dfrac{1}{n}\right)^n$ in that well-known way.

The binomial theorem gives the rule that

$$(a + b)^n = a^n + n\,\frac{a^{n-1}b}{1!} + n(n - 1)\frac{a^{n-2}b^2}{2!}$$

$$+ n(n - 1)(n - 2)\frac{a^{n-3}b^3}{3!} + \ldots .$$

Putting $a = 1$ and $b = \dfrac{1}{n}$, we get

$$\left(1 + \frac{1}{n}\right)^n = 1 + 1 + \frac{1}{2!}\left(\frac{n - 1}{n}\right) + \frac{1}{3!}\frac{(n - 1)(n - 2)}{n^2}$$

$$+ \frac{1}{4!}\frac{(n - 1)(n - 2)(n - 3)}{n^3} + \ldots .$$

Now, if we suppose n to become infinitely great, say a billion, or a billion billions, then $n - 1$, $n - 2$, and $n - 3$. etc., will all be sensibly equal to n; and then the series becomes

$$e = 1 + 1 + \frac{1}{2!} + \frac{1}{3!} + \frac{1}{4!} + \ldots .$$

By taking this rapidly convergent series to as many terms as we please, we can work out the sum to any desired point of accuracy. Here is the working for ten terms:

	1.000000
dividing by 1	1.000000
dividing by 2	0.500000
dividing by 3	0.166667
dividing by 4	0.041667
dividing by 5	0.008333
dividing by 6	0.001389
dividing by 7	0.000198
dividing by 8	0.000025
dividing by 9	0.000003
Total	2.718282

e is incommensurable with 1, and resembles π in being an interminable nonrecurrent decimal.

The Exponential Series.   We shall have need of yet another series. Let us, again making use of the binomial theorem, expand the expression $\left(1 + \dfrac{1}{n}\right)^{nx}$, which is the same as e^x when we make n indefinitely great.

$$e^x = 1^{nx} + nx \; \frac{1^{nx-1}\left(\dfrac{1}{n}\right)}{1!} + nx(nx-1) \; \frac{1^{nx-2}\left(\dfrac{1}{n}\right)^2}{2!}$$

$$+ \; nx(nx-1)(nx-2)\frac{1^{nx-3}\left(\dfrac{1}{n}\right)^3}{3!} + \dots .$$

$$= 1 + x + \frac{1}{2!} \cdot \frac{n^2 x^2 - nx}{n^2} + \frac{1}{3!} \cdot \frac{n^3 x^3 - 3n^2 x^2 + 2nx}{n^3} + \dots$$

$$= 1 + x + \frac{x^2 - \dfrac{x}{n}}{2!} + \frac{x^3 - \dfrac{3x^2}{n} + \dfrac{2x}{n^2}}{3!} + \dots$$

But, when n is made infinitely great, this simplifies down to the following:

$$e^x = 1 + x + \frac{x^2}{2!} + \frac{x^3}{3!} + \frac{x^4}{4!} + \dots$$

This series is called *the exponential series.*

The great reason why e is regarded of importance is that e^x possesses a property, not possessed by any other function of x, that *when you differentiate it its value remains unchanged;* or, in other words, its derivative is the same as itself. This can be instantly seen by differentiating it with respect to x, thus:

$$\frac{d(e^x)}{dx} = 0 + 1 + \frac{2x}{1 \cdot 2} + \frac{3x^2}{1 \cdot 2 \cdot 3} + \frac{4x^3}{1 \cdot 2 \cdot 3 \cdot 4} + \frac{5x^4}{1 \cdot 2 \cdot 3 \cdot 4 \cdot 5} + \dots$$

or $\quad = 1 + x + \dfrac{x^2}{1 \cdot 2} + \dfrac{x^3}{1 \cdot 2 \cdot 3} + \dfrac{x^4}{1 \cdot 2 \cdot 3 \cdot 4} + \dots$

which is exactly the same as the original series.

Now we might have gone to work the other way, and said: Go to; let us find a function of x, such that its derivative is the same as itself. Or, is there any expression, involving only powers of x, which is unchanged by differentiation? Accordingly, let us *assume* as a general expression that

$$y = A + Bx + Cx^2 + Dx^3 + Ex^4 + \dots$$

(in which the coefficients A, B, C, ... will have to be determined), and differentiate it.

$$\frac{dy}{dx} = B + 2Cx + 3Dx^2 + 4Ex^3 + \dots$$

Now, if this new expression is really to be the same as that from which it was derived, it is clear that A *must* $= B$; that $C = \dfrac{B}{2} = \dfrac{A}{1 \cdot 2}$; that $D = \dfrac{C}{3} = \dfrac{A}{1 \cdot 2 \cdot 3}$; that $E = \dfrac{D}{4} = \dfrac{A}{1 \cdot 2 \cdot 3 \cdot 4}$ . . . .

The law of change is therefore that

$$y = A\left(1 + \frac{x}{1} + \frac{x^2}{1 \cdot 2} + \frac{x^3}{1 \cdot 2 \cdot 3} + \frac{x^4}{1 \cdot 2 \cdot 3 \cdot 4} + \ldots\right)$$

If, now, we take $A = 1$ for the sake of further simplicity, we have

$$y = 1 + \frac{x}{1} + \frac{x^2}{1 \cdot 2} + \frac{x^3}{1 \cdot 2 \cdot 3} + \frac{x^4}{1 \cdot 2 \cdot 3 \cdot 4} + \ldots$$

Differentiating it any number of times will give always the same series over again.

If, now, we take the particular case of $A = 1$, and evaluate the series, we shall get simply

when $x = 1$, $y = 2.718281 \ldots$;   that is, $y = e$;

when $x = 2$, $y = (2.718281 \ldots)^2$;   that is, $y = e^2$;

when $x = 3$, $y = (2.718281 \ldots)^3$;   that is, $y = e^3$;

and therefore

when $x = x$, $y = (2.718281 \ldots)^x$;   that is, $y = e^x$,

thus finally demonstrating that

$$e^x = 1 + \frac{x}{1} + \frac{x^2}{1 \cdot 2} + \frac{x^3}{1 \cdot 2 \cdot 3} + \frac{x^4}{1 \cdot 2 \cdot 3 \cdot 4} + \ldots$$

Natural or Napierian Logarithms.

Another reason why e is important is because it was made by Napier, the inventor of logarithms, the basis of his system. If y is the value of e^x, then x is the *logarithm,* to the base e, of y. Or, if

$$y = e^x$$

then $x = \log_e y$

The two curves plotted in Figs. 38 and 39 represent these equations.

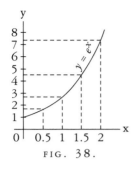

FIG. 38.

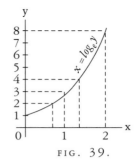

FIG. 39.

The points calculated are:

For Fig. 38

x	0	0.5	1	1.5	2
y	1	1.65	2.72	4.48	7.39

For Fig. 39

y	1	2	3	4	8
x	0	0.69	1.10	1.39	2.08

It will be seen that, though the calculations yield different points for plotting, yet the result is identical. The two equations really mean the same thing.

As many persons who use ordinary logarithms, which are calculated to base 10 instead of base e, are unfamiliar with the "natural" logarithms, it may be worth while to say a word about them.[2] The ordinary rule that adding logarithms gives the logarithm of the product still holds good; or

$$\ln a + \ln b = \ln ab$$

Also the rule of powers holds good;

$$n \times \ln a = \ln a^n$$

But as 10 is no longer the basis, one cannot multiply by 100 or 1000 by merely adding 2 or 3 to the index. A natural logarithm

2. It is customary today to write ln (pronounced el-en) rather than $\log_e$ for all natural logarithms. From here on I have replaced Thompson's $\log_e$ with ln—M.G.

is connected to the common logarithm of the same number by the relations:

$$\log_{10} x = \log_{10} e \times \ln x, \quad \text{and} \quad \ln x = \ln 10 \times \log_{10} x;$$

but $\quad \log_{10} e = \log_{10} 2.718 = 0.4343 \quad$ and $\quad \ln 10 = 2.3026$

$$\log_{10} x = 0.4343 \times \ln x$$

$$\ln x = 2.3026 \times \log_{10} x$$

A Useful Table of "Naperian Logarithms"
(Also called Natural Logarithms or Hyperbolic Logarithms)

Number	Log_e	Number	Log_e
1	0.0000	6	1.7918
1.1	0.0953	7	1.9459
1.2	0.1823	8	2.0794
1.5	0.4055	9	2.1972
1.7	0.5306	10	2.3026
2.0	0.6931	20	2.9957
2.2	0.7885	50	3.9120
2.5	0.9163	100	4.6052
2.7	0.9933	200	5.2983
2.8	1.0296	500	6.2146
3.0	1.0986	1,000	6.9078
3.5	1.2528	2,000	7.6009
4.0	1.3863	5,000	8.5172
4.5	1.5041	10,000	9.2103
5.0	1.6094	20,000	9.9035

Exponential and Logarithmic Equations.
Now let us try our hands at differentiating certain expressions that contain logarithms or exponentials.

Take the equation:

$$y = \ln x$$

First transform this into

$$e^y = x$$

whence, since the derivative of e^y with regard to y is the original function unchanged,

$$\frac{dx}{dy} = e^y$$

and, reverting from the inverse to the original function,

$$\frac{dy}{dx} = \frac{1}{\dfrac{dx}{dy}} = \frac{1}{e^y} = \frac{1}{x}$$

Now this is a very curious result. It may be written

$$\frac{d(\ln x)}{dx} = x^{-1}$$

Note that x^{-1} is a result that we could never have got by the rule for differentiating powers. That rule is to multiply by the power, and reduce the power by 1. Thus, differentiating x^3 gave us $3x^2$; and differentiating x^2 gave $2x^1$. But differentiating x^0 gives us $0 \times x^{-1} = 0$, because x^0 is itself $= 1$, and is a constant. We shall have to come back to this curious fact that differentiating $\log_e x$ gives us $\dfrac{1}{x}$ when we reach the chapter on integrating.

Now, try to differentiate

$$y = \ln (x + a)$$

that is

$$e^y = x + a$$

we have $\dfrac{d(x + a)}{dy} = e^y$, since the derivative of e^y remains e^y.

This gives

$$\frac{dx}{dy} = e^y = x + a;$$

hence, reverting to the original function, we get

$$\frac{dy}{dx} = \frac{1}{\dfrac{dx}{dy}} = \frac{1}{x+a}$$

Next try
$$y = \log_{10} x$$

First change to natural logarithms by multiplying by the modulus 0.4343. This gives us

$$y = 0.4343 \ln x$$

whence
$$\frac{dy}{dx} = \frac{0.4343}{x}$$

The next thing is not quite so simple. Try this:

$$y = a^x$$

Taking the logarithm of both sides, we get

$$\ln y = x \ln a$$

or
$$x = \frac{\ln y}{\ln a} = \frac{1}{\ln a} \times \ln y$$

Since $\dfrac{1}{\ln a}$ is a constant, we get

$$\frac{dx}{dy} = \frac{1}{\ln a} \times \frac{1}{y} = \frac{1}{a^x \times \ln a}$$

hence, reverting to the original function,

$$\frac{dy}{dx} = \frac{1}{\dfrac{dx}{dy}} = a^x \times \ln a$$

We see that, since

$$\frac{dx}{dy} \times \frac{dy}{dx} = 1 \quad \text{and} \quad \frac{dx}{dy} = \frac{1}{y} \times \frac{1}{\ln a}, \quad \frac{1}{y} \times \frac{dy}{dx} = \ln a$$

We shall find that whenever we have an expression such as $\ln y =$ a function of x, we always have $\dfrac{1}{y}\dfrac{dy}{dx} =$ the derivative of the function of x, so that we could have written at once, from $\ln y = x \ln a$

$$\frac{1}{y}\frac{dy}{dx} = \ln a \quad \text{and} \quad \frac{dy}{dx} = y \ln a = a^x \ln a$$

Let us now attempt further examples.

Examples.
(1) $y = e^{-ax}$. Let $z = -ax$; then $y = e^z$

$$\frac{dy}{dz} = e^z; \quad \frac{dz}{dx} = -a; \text{ hence } \frac{dy}{dx} = -ae^z = -ae^{-ax}$$

Or thus:

$$\ln y = -ax; \quad \frac{1}{y}\frac{dy}{dx} = -a; \quad \frac{dy}{dx} = -ay = -ae^{-ax}$$

(2) $y = e^{\frac{x^2}{3}}$. Let $z = \dfrac{x^2}{3}$; then $y = e^z$.

$$\frac{dy}{dz} = e^z; \quad \frac{dz}{dx} = \frac{2x}{3}; \quad \frac{dy}{dx} = \frac{2x}{3}e^{\frac{x^2}{3}}$$

Or thus: $\quad \ln y = \dfrac{x^2}{3}; \quad \dfrac{1}{y}\dfrac{dy}{dx} = \dfrac{2x}{3}; \quad \dfrac{dy}{dx} = \dfrac{2x}{3}e^{\frac{x^2}{3}}$

(3) $y = e^{\frac{2x}{x+1}}$. $\ln y = \dfrac{2x}{x+1}$, $\dfrac{1}{y}\dfrac{dy}{dx} = \dfrac{2(x+1) - 2x}{(x+1)^2}$

hence $\qquad \dfrac{dy}{dx} = \dfrac{2y}{(x+1)^2} = \dfrac{2}{(x+1)^2}e^{\frac{2x}{x+1}}$

Check by writing $z = \dfrac{2x}{x+1}$

(4) $y = e^{\sqrt{x^2 + a}}$. $\ln y = (x^2 + a)^{\frac{1}{2}}$

$$\frac{1}{y}\frac{dy}{dx} = \frac{x}{(x^2 + a)^{\frac{1}{2}}} \quad \text{and} \quad \frac{dy}{dx} = \frac{x \times e^{\sqrt{x^2 + a}}}{(x^2 + a)^{\frac{1}{2}}}$$

For if $u = (x^2 + a)^{\frac{1}{2}}$ and $v = x^2 + a$, $u = v^{\frac{1}{2}}$,

$$\frac{du}{dv} = \frac{1}{2v^{\frac{1}{2}}}; \frac{dv}{dx} = 2x; \frac{du}{dx} = \frac{x}{(x^2 + a)^{\frac{1}{2}}}$$

Check by writing $z = \sqrt{x^2 + a}$

(5) $y = \ln (a + x^3)$. Let $z = (a + x^3)$; then $y = \ln z$.

$$\frac{dy}{dz} = \frac{1}{z}; \frac{dz}{dx} = 3x^2; \text{hence } \frac{dy}{dx} = \frac{3x^2}{a + x^3}$$

(6) $y = \ln\{3x^2 + \sqrt{a + x^2}\}$.   Let $z = 3x^2 + \sqrt{a + x^2}$;   then $y = \ln z$.

$$\frac{dy}{dz} = \frac{1}{z}; \frac{dz}{dx} = 6x + \frac{x}{\sqrt{x^2 + a}}$$

$$\frac{dy}{dx} = \frac{6x + \dfrac{x}{\sqrt{x^2 + a}}}{3x^2 + \sqrt{a + x^2}} = \frac{x\left(1 + 6\sqrt{x^2 + a}\right)}{\left(3x^2 + \sqrt{x^2 + a}\right)\sqrt{x^2 + a}}$$

(7) $y = (x + 3)^2 \sqrt{x - 2}$

$$\ln y = 2 \ln (x + 3) + \tfrac{1}{2} \ln (x - 2)$$

$$\frac{1}{y}\frac{dy}{dx} = \frac{2}{(x + 3)} + \frac{1}{2(x - 2)}$$

$$\frac{dy}{dx} = (x + 3)^2\sqrt{x - 2}\left\{\frac{2}{x + 3} + \frac{1}{2(x - 2)}\right\} = \frac{5(x + 3)(x - 1)}{2\sqrt{x - 2}}$$

(8) $y = (x^2 + 3)^3(x^3 - 2)^{\frac{2}{3}}$.

$$\ln y = 3 \ln (x^2 + 3) + \tfrac{2}{3} \ln (x^3 - 2)$$

$$\frac{1}{y}\frac{dy}{dx} = 3\frac{2x}{x^2 + 3} + \frac{2}{3}\frac{3x^2}{x^3 - 2} = \frac{6x}{x^2 + 3} + \frac{2x^2}{x^3 - 2}$$

For if $u = \ln (x^2 + 3)$, $z = x^2 + 3$ and $u = \ln z$

$$\frac{du}{dz} = \frac{1}{z}; \frac{dz}{dx} = 2x; \frac{du}{dx} = \frac{2x}{z} = \frac{2x}{x^2 + 3}$$

Similarly, if $v = \ln (x^3 - 2)$, $\dfrac{dv}{dx} = \dfrac{3x^2}{x^3 - 2}$ and

$$\frac{dy}{dx} = (x^2 + 3)^3 (x^3 - 2)^{\frac{2}{3}} \left\{ \frac{6x}{x^2 + 3} + \frac{2x^2}{x^3 - 2} \right\}$$

(9) $y = \dfrac{\sqrt[2]{x^2 + a}}{\sqrt[3]{x^3 - a}}$

$$\ln y = \frac{1}{2} \ln (x^2 + a) - \frac{1}{3} \ln (x^3 - a)$$

$$\frac{1}{y} \frac{dy}{dx} = \frac{1}{2} \frac{2x}{x^2 + a} - \frac{1}{3} \frac{3x^2}{x^3 - a} = \frac{x}{x^2 + a} - \frac{x^2}{x^3 - a}$$

and $$\frac{dy}{dx} = \frac{\sqrt[2]{x^2 + a}}{\sqrt[3]{x^3 - a}} \left\{ \frac{x}{x^2 + a} - \frac{x^2}{x^3 - a} \right\}$$

(10) $y = \dfrac{1}{\ln x}$.

$$\frac{dy}{dx} = \frac{\ln x \times 0 - 1 \times \dfrac{1}{x}}{\ln^2 x} = -\frac{1}{x \ln^2 x}$$

(11) $y = \sqrt[3]{\ln x} = (\ln x)^{\frac{1}{3}}$. Let $z = \ln x$; $y = z^{\frac{1}{3}}$,

$$\frac{dy}{dz} = \frac{1}{3} z^{-\frac{2}{3}}; \frac{dz}{dx} = \frac{1}{x}; \frac{dy}{dx} = \frac{1}{3x\sqrt[3]{\ln^2 x}}$$

(12) $y = \left(\dfrac{1}{a^x} \right)^{ax}$

$$\ln y = -ax \ln a^x = -ax^2 \cdot \ln a$$

$$\frac{1}{y}\frac{dy}{dx} = -2ax \cdot \ln a$$

and
$$\frac{dy}{dx} = -2ax\left(\frac{1}{a^x}\right)^{ax} \cdot \ln a = -2xa^{1-ax^2} \cdot \ln a$$

Try now the following exercises.

EXERCISES XII

(1) Differentiate $y = b(e^{ax} - e^{-ax})$.

(2) Find the derivative with respect to t of the expression $u = at^2 + 2 \ln t$.

(3) If $y = n^t$, find $\dfrac{d(\ln y)}{dt}$

(4) Show that if $y = \dfrac{1}{b} \cdot \dfrac{a^{bx}}{\ln a}$; $\dfrac{dy}{dx} = a^{bx}$

(5) If $w = pv^n$, find $\dfrac{dw}{dv}$

Differentiate

(6) $y = \ln x^n$ (7) $y = 3e^{-\frac{x}{x-1}}$ (8) $y = (3x^2 + 1)e^{-5x}$

(9) $y = \ln (x^a + a)$ (10) $y = (3x^2 - 1)(\sqrt{x} + 1)$

(11) $y = \dfrac{\ln (x + 3)}{x + 3}$ (12) $y = a^x \times x^a$

(13) It was shown by Lord Kelvin that the speed of signalling through a submarine cable depends on the value of the ratio of the external diameter of the core to the diameter of the enclosed copper wire. If this ratio is called y, then the number of signals s that can be sent per minute can be expressed by the formula

$$s = ay^2 \ln\frac{1}{y}$$

where a is a constant depending on the length and the quality of the materials. Show that if these are given, s will be a maximum if $1/e^{\frac{1}{2}}$.

(14) Find the maximum or minimum of

$$y = x^3 - \ln x$$

(15) Differentiate $y = \ln(axe^x)$

(16) Differentiate $y = (\ln ax)^3$

The Logarithmic Curve

Let us return to the curve which has its successive ordinates in geometrical progression, such as that represented by the equation $y = bp^x$.

We can see, by putting $x = 0$, that b is the initial height of y. Then when

$$x = 1, y = bp; \quad x = 2, y = bp^2; \quad x = 3, y = bp^3, \text{ etc.}$$

Also, we see that p is the numerical value of the ratio between the height of any ordinate and that of the next preceding it. In Fig. 40, we have taken p as $\frac{6}{5}$; each ordinate being $\frac{6}{5}$ as high as the preceding one.

If two successive ordinates are related together thus in a constant ratio, their logarithms will have a constant difference; so that, if we should plot out a new curve, Fig. 41, with values of $\ln y$ as ordinates, it would be a straight line sloping up by equal steps. In fact, it follows from the equation, that

$$\ln y = \ln b + x \cdot \ln p, \text{ whence } \ln y - \ln b = x \cdot \ln p.$$

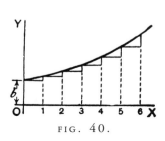

FIG. 40.

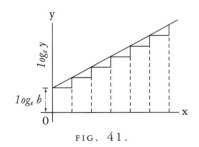

FIG. 41.

Now, since $\ln p$ is a mere number, and may be written as $\ln p = a$, it follows that

$$\ln \frac{y}{b} = ax$$

and the equation takes the new form

$$y = be^{ax}$$

The Die-away Curve

If we were to take p as a proper fraction (less than unity), the curve would obviously tend to sink downwards, as in Fig. 42, where each successive ordinate is $\frac{3}{4}$ of the height of the preceding one.

The equation is still

$$y = bp^x$$

but since p is less than one, $\ln p$ will be a negative quantity, and may be written $-a$; so that $p = e^{-a}$, and now our equation for the curve takes the form

$$y = be^{-ax}$$

The importance of this expression is that, in the case where the independent variable is *time,* the equation represents the course of a great many physical processes in which something is *gradually dying away.* Thus, the cooling of a hot body is represented (in Newton's celebrated "law of cooling") by the equation

$$\theta_t = \theta_0 e^{-at}$$

where θ_0 is the original excess of temperature of a hot body over that of its surroundings, θ_t the excess of temperature at the end of time t, and a is a constant—namely, the constant of decrement, depending on the amount of surface exposed by the body, and on its coefficients of conductivity and emissivity, etc.

A similar formula,

$$Q_t = Q_0 e^{-at}$$

is used to express the charge of an electrified body, originally having a charge Q_0, which is leaking away with a con

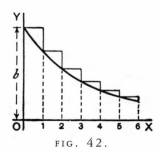

FIG. 42.

stant of decrement a; which constant depends in this case on the capacity of the body and on the resistance of the leakage-path.

Oscillations given to a flexible spring die out after a time, and the dying-out of the amplitude of the motion may be expressed in a similar way.

In fact e^{-at} serves as a *die-away factor* for all those phenomena in which the rate of decrease is proportional to the magnitude of that which is decreasing; or where, in our usual symbols, $\dfrac{dy}{dt}$ is proportional at every moment to the value that y has at that moment. For we have only to inspect the curve, Fig. 42, to see that at every part of it, the slope $\dfrac{dy}{dx}$ is proportional to the height y; the curve becoming flatter as y grows smaller. In symbols, thus

$$y = be^{-ax}$$

or
$$\ln y = \ln b - ax \ln e = \ln b - ax,$$

and, differentiating,
$$\frac{1}{y} \frac{dy}{dx} = -a$$

hence,
$$\frac{dy}{dx} = be^{-ax} \times (-a) = -ay$$

or, in words, the slope of the curve is downward, and proportional to y and to the constant a.

We should have got the same result if we had taken the equation in the form

$$y = bp^x$$

for then
$$\frac{dy}{dx} = bp^x \times \ln p$$

But
$$\ln p = -a$$

giving us
$$\frac{dy}{dx} = y \times (-a) = -ay$$

as before.

The Time-Constant.   In the expression for the "die-away factor" e^{-at}, the quantity a is the reciprocal of another quantity known as *"the time-constant"*, which we may denote by the symbol T. Then the die-away factor will be written $e^{-\frac{t}{T}}$; and it will be seen, by making $t = T$, that the meaning of T $\left(\text{or of } \dfrac{1}{a}\right)$ is that this is the length of time which it takes for the original quantity (called θ_0 or Q_0 in the preceding instances) to die away to $\dfrac{1}{e}$-th part—that is to 0.3679—of its original value.

The values of e^x and e^{-x} are continually required in different branches of physics, and as they are given in very few sets of mathematical tables, some of the values are tabulated here for convenience.

x	e^x	e^{-x}	$1 - e^{-x}$
0.00	1.0000	1.0000	0.0000
0.10	1.1052	0.9048	0.0952
0.20	1.2214	0.8187	0.1813
0.50	1.6487	0.6065	0.3935
0.75	2.1170	0.4724	0.5276
0.90	2.4596	0.4066	0.5934
1.00	2.7183	0.3679	0.6321
1.10	3.0042	0.3329	0.6671
1.20	3.3201	0.3012	0.6988
1.25	3.4903	0.2865	0.7135
1.50	4.4817	0.2231	0.7769
1.75	5.755	0.1738	0.8262
2.00	7.389	0.1353	0.8647
2.50	12.182	0.0821	0.9179
3.00	20.086	0.0498	0.9502
3.50	33.115	0.0302	0.9698
4.00	54.598	0.0183	0.9817
4.50	90.017	0.0111	0.9889
5.00	148.41	0.0067	0.9933
5.50	244.69	0.0041	0.9959
6.00	403.43	0.00248	0.99752
7.50	1808.04	0.00055	0.99945
10.00	22026.5	0.000045	0.999955

As an example of the use of this table, suppose there is a hot body cooling, and that at the beginning of the experiment (*i.e.* when $t = 0$) it is 72° hotter than the surrounding objects, and if the time-constant of its cooling is 20 minutes (that is, if it takes 20 minutes for its excess of temperature to fall to $\dfrac{1}{e}$ part of 72°), then we can calculate to what it will have fallen in any given time t. For instance, let t be 60 minutes. Then $\dfrac{t}{T} = 60 \div 20 = 3$, and we shall have to find the value of e^{-3}, and then multiply the original 72° by this. The table shows that e^{-3} is 0.0498. So that at the end of 60 minutes the excess of temperature will have fallen to $72° \times 0.0498 = 3.586°$.

Further Examples.
(1) The strength of an electric current in a conductor at a time t secs. after the application of the electromotive force producing it is given by the expression $C = \dfrac{E}{R}\left\{1 - e^{-\frac{Rt}{L}}\right\}$.

The time constant is $\dfrac{L}{R}$.

If $E = 10$, $R = 1$, $L = 0.01$; then when t is very large the term $1 - e^{-\frac{Rt}{L}}$ becomes 1, and $C = \dfrac{E}{R} = 10$; also

$$\frac{L}{R} = T = 0.01$$

Its value at any time may be written:
$$C = 10 - 10e^{-\frac{t}{0.01}}$$

the time-constant being 0.01. This means that it takes 0.01 sec. for the variable term to fall to $\dfrac{1}{e} = 0.3679$ of its initial value $10e^{-\frac{0}{0.01}} = 10$.

To find the value of the current when $t = 0.001$ sec., say,
$\dfrac{t}{T} = 0.1$, $e^{-0.1} = 0.9048$ (from table).

It follows that, after 0.001 sec., the variable term is

$$0.9048 \times 10 = 9.048$$

and the actual current is $10 - 9.048 = 0.952$.

Similarly, at the end of 0.1 sec.,

$$\frac{t}{T} = 10; \ e^{-10} = 0.000045$$

the variable term is $10 \times 0.000045 = 0.00045$, the current being 9.9995.

(2) The intensity I of a beam of light which has passed through a thickness l cm. of some transparent medium is $I = I_0 e^{-Kl}$, where I_0 is the initial intensity of the beam and K is a "constant of absorption".

This constant is usually found by experiments. If it be found, for instance, that a beam of light has its intensity diminished by 18% in passing through 10 cm. of a certain transparent medium, this means that $82 = 100 \times e^{-K \times 10}$ or $e^{-10K} = 0.82$, and from the table one sees that $10K = 0.20$ very nearly; hence $K = 0.02$.

To find the thickness that will reduce the intensity to half its value, one must find the value of l which satisfies the equality $50 = 100 \times e^{-0.02l}$, or $0.5 = e^{-0.02l}$. It is found by putting this equation in its natural logarithmic form, namely,

$$l = \frac{\ln 0.5}{-0.02} = 34.7 \text{ cm. nearly.}$$

(3) The quantity Q of a radio-active substance which has not yet undergone transformation is known to be related to the initial quantity Q_0 of the substance by the relation $Q = Q_0 e^{-\lambda t}$, where λ is a constant and t the time in seconds elapsed since the transformation began.

For "Radium A", if time is expressed in seconds, experiment shows that $\lambda = 3.85 \times 10^{-3}$. Find the time required for transforming half the substance. (This time is called the "half life" of the substance.)

We have $$0.5 = e^{-0.00385t}.$$

$$\log_{10} 0.5 = -0.00385t \times \log_{10} e$$

and $$t = 3 \text{ minutes very nearly.}$$

EXERCISES XIII

(1) Draw the curve $y = be^{-\frac{t}{T}}$; where $b = 12$, $T = 8$, and t is given various values from 0 to 20.

(2) If a hot body cools so that in 24 minutes its excess of temperature has fallen to half the initial amount, deduce the time-constant, and find how long it will be in cooling down to 1 percent of the original excess.

(3) Plot the curve $y = 100(1 - e^{-2t})$.

(4) The following equations give very similar curves:

$$(\text{i}) \ y = \frac{ax}{x + b}; \quad (\text{ii}) \ y = a\left(1 - e^{-\frac{x}{b}}\right)$$

$$(\text{iii}) \ y = \frac{a}{90°} \arctan\left(\frac{x}{b}\right)$$

Draw all three curves, taking $a = 100$ millimetres; $b = 30$ millimetres.

(5) Find the derivative of y with respect to x, if

$$(a) \ y = x^x; \quad (b) \ y = (e^x)^x; \quad (c) \ y = e^{x^x}$$

(6) For "Thorium A", the value of λ is 5; find the "half life", that is, the time taken by the transformation of a quantity Q of "Thorium A" equal to half the initial quantity Q_0 in the expression

$$Q = Q_0 e^{-\lambda t};$$

t being in seconds.

(7) A condenser of capacity $K = 4 \times 10^{-6}$, charged to a potential $V_0 = 20$, is discharging through a resistance of 10,000 ohms. Find the potential V after (a) 0.1 second; (b) 0.01 second; assuming that the fall of potential follows the rule $V = V_0 e^{-\frac{t}{KR}}$.

(8) The charge Q of an electrified insulated metal sphere is reduced from 20 to 16 units in 10 minutes. Find the coefficient μ of leakage, if $Q = Q_0 \times e^{-\mu t}$; Q_0 being the initial charge and t being in seconds. Hence find the time taken by half the charge to leak away.

(9) The damping on a telephone line can be ascertained from the relation $i = i_0 e^{-\beta l}$, where i is the strength, after t seconds, of a telephonic current of initial strength i_0; l is the length of the line in kilometres, and β is a constant. For the Franco-English submarine cable laid in 1910, $\beta = 0.0114$. Find the damping at the end of the cable (40 kilometres), and the length along which i is still 8% of the original current (limiting value of very good audition).

(10) The pressure p of the atmosphere at an altitude h kilometres is approximately $p = p_0 \, e^{-kh}$ for some constant k; p_0 being the pressure at sea level (760 millimetres).

The pressures at 10, 20, and 50 kilometres being 199.2, 42.4, 0.32 millimetres respectively, find k in each case. Using the average of these three values of k, find the percentage error in the computed value of the pressure at the three heights.

(11) Find the minimum or maximum of $y = x^x$.

(12) Find the minimum or maximum of $y = x^{\frac{1}{x}}$.

(13) Find the minimum or maximum of $y = xa^{\frac{1}{x}}$, if $a > 1$.

Chapter XV

HOW TO DEAL WITH SINES AND COSINES

———

Greek letters being usual to denote angles, we will take as the usual letter for any variable angle the letter θ ("theta").

Let us consider the function

$$y = \sin \theta$$

What we have to investigate is the value of $\dfrac{d(\sin \theta)}{d\theta}$; or, in other words, if the angle θ varies, we have to find the relation between the increment of the sine and the increment of the angle, both increments being infinitely small in themselves. Examine Fig. 43, wherein, if the radius of the circle is unity, the height of y is the sine, and θ is the angle. Now, if θ is supposed to increase by the addition to it of the small angle $d\theta$—an element of angle—the height of y, the sine, will be increased by a small element dy. The new height $y + dy$ will be the sine of the new angle $\theta + d\theta$, or, stating it as an equation,

$$y + dy = \sin (\theta + d\theta)$$

and subtracting from this the first equation gives

$$dy = \sin (\theta + d\theta) - \sin \theta$$

The quantity on the right-hand side is the difference between two sines, and books on trigonometry tell us how to work this out. For they tell us that if M and N are two different angles,

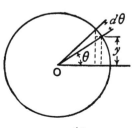

FIG. 43.

175

$$\sin M - \sin N = 2 \cos \frac{M+N}{2} \cdot \sin \frac{M-N}{2}$$

If, then, we put $M = \theta + d\theta$ for one angle, and $N = \theta$ for the other, we may write

$$dy = 2 \cos \frac{\theta + d\theta + \theta}{2} \cdot \sin \frac{\theta + d\theta - \theta}{2}$$

or, $$dy = 2 \cos (\theta + \tfrac{1}{2}d\theta) \cdot \sin \tfrac{1}{2}d\theta$$

But if we regard $d\theta$ as infinitely small, then in the limit we may neglect $\tfrac{1}{2}d\theta$ by comparison with θ, and may also take $\sin \tfrac{1}{2}d\theta$ as being the same as $\tfrac{1}{2}d\theta$. The equation then becomes:

$$dy = 2 \cos \theta \cdot \tfrac{1}{2}d\theta$$

$$dy = \cos \theta \cdot d\theta$$

and, finally, $$\frac{dy}{d\theta} = \cos \theta$$

The accompanying curves, Figs. 44 and 45, show, plotted to scale, the values of $y = \sin \theta$, and $\dfrac{dy}{d\theta} = \cos \theta$, for the corresponding values of θ.

Take next the cosine.

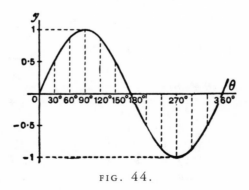

FIG. 44.

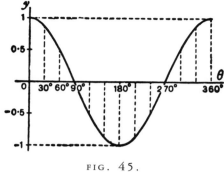

FIG. 45.

Let $\quad y = \cos \theta$

Now $\quad \cos \theta = \sin \left(\dfrac{\pi}{2} - \theta \right)$

Therefore

$$dy = d \left(\sin \left(\dfrac{\pi}{2} - \theta \right) \right) = \cos \left(\dfrac{\pi}{2} - \theta \right) \times d(-\theta)$$

$$= \cos \left(\dfrac{\pi}{2} - \theta \right) \times (-d\theta)$$

$$\dfrac{dy}{d\theta} = -\cos \left(\dfrac{\pi}{2} - \theta \right)$$

And it follows that

$$\dfrac{dy}{d\theta} = -\sin \theta$$

Lastly, take the tangent.

Let

$$y = \tan \theta$$

$$= \dfrac{\sin \theta}{\cos \theta}$$

Applying the rule given in Chapter VI for differentiating a quotient of two functions, we get

$$\frac{dy}{d\theta} = \frac{\cos\theta\,\dfrac{d(\sin\theta)}{d\theta} - \sin\theta\,\dfrac{d(\cos\theta)}{d\theta}}{\cos^2\theta}$$

$$= \frac{\cos^2\theta + \sin^2\theta}{\cos^2\theta}$$

$$= \frac{1}{\cos^2\theta}$$

or $\quad \dfrac{dy}{d\theta} = \sec^2\theta$

Collecting these results, we have:

y	$\dfrac{dy}{d\theta}$
$\sin\theta$	$\cos\theta$
$\cos\theta$	$-\sin\theta$
$\tan\theta$	$\sec^2\theta$

Sometimes, in mechanical and physical questions, as, for example, in simple harmonic motion and in wave-motions, we have to deal with angles that increase in proportion to the time. Thus, if T be the time of one complete *period*, or movement round the circle, then, since the angle all round the circle is 2π radians, or $360°$, the amount of angle moved through in time t will be

$$\theta = 2\pi\,\frac{t}{T}, \text{ in radians}$$

or $\quad \theta = 360\dfrac{t}{T}, \text{ in degrees}$

If the *frequency*, or number of periods per second, be denoted by n, then $n = \dfrac{1}{T}$, and we may then write:

$$\theta = 2\pi nt$$

Then we shall have $\quad y = \sin 2\pi nt$

If, now, we wish to know how the sine varies with respect to time, we must differentiate with respect, not to θ, but to t. For this we must resort to the artifice explained in Chapter IX, and put

$$\frac{dy}{dt} = \frac{dy}{d\theta} \cdot \frac{d\theta}{dt}$$

Now $\dfrac{d\theta}{dt}$ will obviously be $2\pi n$; so that

$$\frac{dy}{dt} = \cos\,\theta \times 2\pi n$$

$$= 2\pi n \cdot \cos\,2\pi nt$$

Similarly, it follows that

$$\frac{d(\cos 2\pi nt)}{dt} = -2\pi n \cdot \sin 2\pi nt$$

Second Derivative of Sine or Cosine

We have seen that when $\sin\,\theta$ is differentiated with respect to θ it becomes $\cos\,\theta$; and that when $\cos\,\theta$ is differentiated with respect to θ it becomes $-\sin\,\theta$; or, in symbols,

$$\frac{d^2(\sin\,\theta)}{d\theta^2} = -\sin\,\theta$$

So we have this curious result that we have found a function such that if we differentiate it twice over, we get the same thing from which we started, but with the sign changed from $+$ to $-$.

The same thing is true for the cosine; for differentiating $\cos\,\theta$ gives us $-\sin\,\theta$, and differentiating $-\sin\,\theta$ gives us $-\cos\,\theta$; or thus:

$$\frac{d^2(\cos\,\theta)}{d\theta^2} = -\cos\,\theta.$$

Sines and cosines furnish a basis for the only functions of which the second derivative is equal and of opposite sign to the original function.

Examples.
With what we have so far learned we can now differentiate expressions of a more complex nature.

(1) $y = \text{arcsin } x$.

If y is the angle whose sine is x, then $x = \sin y$.

$$\frac{dx}{dy} = \cos y$$

Passing now from the inverse function to the original one, we get

$$\frac{dy}{dx} = \frac{1}{\dfrac{dx}{dy}} = \frac{1}{\cos y}$$

Now
$$\cos y = \sqrt{1 - \sin^2 y} = \sqrt{1 - x^2}$$

hence
$$\frac{dy}{dx} = \frac{1}{\sqrt{1 - x^2}}$$

a rather unexpected result. Because by definition, $-\dfrac{\pi}{2} \leq \text{arcsin } y \leq \dfrac{\pi}{2}$, we know that $\cos y$ is positive, so we use the positive square root here.

(2) $y = \cos^3 \theta$.

This is the same thing as $y = (\cos \theta)^3$

Let $v = \cos \theta$; then $y = v^3$; $\dfrac{dy}{dv} = 3v^2$

$$\frac{dv}{d\theta} = -\sin \theta$$

$$\frac{dy}{d\theta} = \frac{dy}{dv} \times \frac{dv}{d\theta} = -3 \cos^2\theta \sin \theta$$

(3) $y = \sin (x + a)$.

Let $v = x + a$; then $y = \sin v$.

$$\frac{dv}{dx} = 1; \quad \frac{dy}{dv} = \cos v \quad \text{and} \quad \frac{dy}{dx} = \cos (x + a)$$

(4) $y = log_e \sin \theta$

Let $v = \sin \theta$; $y = log_e v$.

$$\frac{dv}{d\theta} = \cos \theta; \quad \frac{dy}{dv} = \frac{1}{v}; \quad \frac{dy}{d\theta} = \frac{1}{\sin \theta} \times \cos \theta = \cot \theta.$$

(5) $y = \cot \theta = \dfrac{\cos \theta}{\sin \theta}$

$$\frac{dy}{d\theta} = \frac{-\sin^2\theta - \cos^2\theta}{\sin^2\theta}$$

$$= -(1 + \cot^2\theta) = -\csc^2\theta$$

(6) $y = \tan 3\theta$

Let $v = 3\theta$; $y = \tan v$; $\dfrac{dv}{d\theta} = 3$; $\dfrac{dy}{dv} = \sec^2 v$

and

$$\frac{dy}{d\theta} = 3 \sec^2 3\theta$$

(7) $y = \sqrt{1 + 3 \tan^2\theta} = (1 + 3 \tan^2\theta)^{\frac{1}{2}}$

Let $v = 3 \tan^2\theta$, then $y = (1 + v)^{\frac{1}{2}}$;

$$\frac{dv}{d\theta} = 6 \tan \theta \sec^2\theta; \quad \frac{dy}{dv} = \frac{1}{2\sqrt{1 + v}}$$

and

$$\frac{dy}{d\theta} = \frac{6 \tan \theta \sec^2\theta}{2\sqrt{1 + v}} = \frac{6 \tan \theta \sec^2\theta}{2\sqrt{1 + 3 \tan^2\theta}}$$

for, if $u = \tan \theta$, $v = 3u^2$

$$\frac{du}{d\theta} = \sec^2\theta; \quad \frac{dv}{du} = 6u$$

hence
$$\frac{dv}{d\theta} = 6 \tan \theta \sec^2\theta$$

hence
$$\frac{dy}{d\theta} = \frac{6 \tan \theta \sec^2\theta}{2\sqrt{1 + 3 \tan^2\theta}}$$

(8) $y = \sin x \cos x$.

$$\frac{dy}{dx} = \sin x(-\sin x) + \cos x \times \cos x$$

$$= \cos^2 x - \sin^2 x$$

Closely associated with *sines, cosines* and *tangents* are three other very useful functions. They are the hyperbolic sine, cosine and tangent, and are written *sinh, cosh, tanh*. These functions are defined as follows:

$$\sinh x = \tfrac{1}{2}(e^x - e^{-x}), \quad \cosh x = \tfrac{1}{2}(e^x + e^{-x}),$$

$$\tanh x = \frac{\sinh x}{\cosh x} = \frac{e^x - e^{-x}}{e^x + e^{-x}}.$$

Between $\sinh x$ and $\cosh x$, there is an important relation, for

$$\cosh^2 x - \sinh^2 x = \tfrac{1}{4}(e^x + e^{-x})^2 - \tfrac{1}{4}(e^x - e^{-x})^2$$

$$= \tfrac{1}{4}(e^{2x} + 2 + e^{-2x} - e^{2x} + 2 - e^{-2x}) = 1.$$

Now
$$\frac{d}{dx}(\sinh x) = \tfrac{1}{2}(e^x + e^{-x}) = \cosh x.$$

$$\frac{d}{dx}(\cosh x) = \tfrac{1}{2}(e^x - e^{-x}) = \sinh x.$$

$$\frac{d}{dx}(\tanh x) = \frac{\cosh x \dfrac{d}{dx}(\sinh x) - \sinh x \dfrac{d}{dx}(\cosh x)}{\cosh^2 x}$$

$$= \frac{\cosh^2 x - \sinh^2 x}{\cosh^2 x} = \frac{1}{\cosh^2 x}$$

by the relation just proved.

EXERCISES XIV

(1) Differentiate the following:

$$\text{(i) } y = A \sin \left(\theta - \frac{\pi}{2} \right)$$

$$\text{(ii) } y = \sin^2 \theta; \text{ and } y = \sin 2\theta$$

$$\text{(iii) } y = \sin^3 \theta; \text{ and } y = \sin 3\theta$$

(2) Find the value of θ $(0 \leq \theta \leq 2\pi)$ for which $\sin \theta \times \cos \theta$ is a maximum.

(3) Differentiate $y = \dfrac{1}{2\pi} \cos 2\pi nt$.

(4) If $y = \sin a^x$, find $\dfrac{dy}{dx}$.   (5) Differentiate $y = \ln \cos x$.

(6) Differentiate $y = 18.2 \sin (x + 26°)$.

(7) Plot the curve $y = 100 \sin (\theta - 15°)$; and show that the slope of the curve at $\theta = 75°$ is half the maximum slope.

(8) If $y = \sin \theta \cdot \sin 2\theta$, find $\dfrac{dy}{d\theta}$.

(9) If $y = a \cdot \tan^m(\theta^n)$, find the derivative of y with respect to θ.

(10) Differentiate $y = e^x \sin^2 x$.

(11) Differentiate the three equations of Exercises XIII, No. 4, and compare their derivatives, as to whether they are equal, or nearly equal, for very small values of x, or for very large values of x, or for values of x in the neighbourhood of $x = 30$.

(12) Differentiate the following:

(i) $y = \sec x$.   (ii) $y = \arccos x$.

(iii) $y = \arctan x$.   (iv) $y = \text{arcsec } x$.

(v) $y = \tan x \times \sqrt{3} \sec x$.

(13) Differentiate $y = \sin (2\theta + 3)^{2.3}$.

(14) Differentiate $y = \theta^3 + 3 \sin (\theta + 3) - 3^{\sin\theta} - 3^\theta$.

(15) Find the maximum or minimum of $y = \theta \cos \theta$, for $-\dfrac{\pi}{2} \leq \theta \leq \dfrac{\pi}{2}$.

Chapter XVI

PARTIAL DIFFERENTIATION

———∞∞∞———

We sometimes come across quantities that are functions of more than one independent variable. Thus, we may find a case where y depends on two other variable quantities, one of which we will call u and the other v. In symbols

$$y = f(u, v)$$

Take the simplest concrete case.

Let $\qquad\qquad y = u \times v$

What are we to do? If we were to treat v as a constant, and differentiate with respect to u, we should get

$$dy_v = v \, du$$

or if we treat u as a constant, and differentiate with respect to v, we should have:

$$dy_u = u \, dv$$

The little letters here put as subscripts are to show which quantity has been taken as constant in the operation.

Another way of indicating that the differentiation has been performed only *partially,* that is, has been performed only with-respect to *one* of the independent variables, is to write the derivatives with a symbol based on Greek small deltas, instead of little d.[1] In this way

$$\frac{\partial y}{\partial u} = v$$

$$\frac{\partial y}{\partial v} = u$$

If we put in these values for v and u respectively, we shall have

$$\left. \begin{array}{l} dy_v = \dfrac{\partial y}{\partial u} du, \\[2em] dy_u = \dfrac{\partial y}{\partial v} dv, \end{array} \right\} \text{ which are } \textit{partial derivatives.}$$

But, if you think of it, you will observe that the total variation of y depends on *both* these things at the same time. That is to say, if both are varying, the real dy ought to be written

$$dy = \frac{\partial y}{\partial u} du + \frac{\partial y}{\partial v} dv$$

and this is called a *total differential.* In some books it is written

$$dy = \left(\frac{dy}{du} \right) du + \left(\frac{dy}{dv} \right) dv.$$

---

1. Robert Ainsley, in his amusing booklet *Bluff Your Way in Mathematics* (1988) defines partial derivatives as "derivatives biased toward x, y, or z instead of treating all three equally—the sign for this is six written backwards . . . ."

The adjective "partial" indicates that the derivative is partial toward one independent variable, the other or others being treated as constants. A derivative is called a "mixed partial derivative" if it is a partial derivative of order 2 or higher that involves more than one of the independent variables. Higher partial derivatives are partial derivatives of partial derivatives.—M.G.

Example (1).   Find the partial derivatives of the expression $w = 2ax^2 + 3bxy + 4cy^3$. The answers are:

$$\left.\begin{array}{l} \dfrac{\partial w}{\partial x} = 4ax + 3by \\[3ex] \dfrac{\partial w}{\partial y} = 3bx + 12cy^2 \end{array}\right\}$$

The first is obtained by supposing y constant, the second is obtained by supposing x constant; then the total differential is

$$dw = (4ax + 3by)dx + (3bx + 12cy^2)dy$$

Example (2).   Let $z = x^y$. Then, treating first y and then x as constant, we get in the usual way

$$\left.\begin{array}{l} \dfrac{\partial z}{\partial x} = yx^{y-1} \\[3ex] \dfrac{\partial z}{\partial y} = x^y \times log_e\, x \end{array}\right\}$$

so that $dz = yx^{y-1}dx + x^y\, log_e\, xdy$.

Example (3).   A cone having height h and radius of base r, has volume $V = \frac{1}{3}\pi r^2 h$. If its height remains constant, while r changes, the ratio of change of volume, with respect to radius, is different from ratio of change of volume with respect to height which would occur if the height were varied and the radius kept constant, for

$$\left.\begin{array}{l} \dfrac{\partial V}{\partial r} = \dfrac{2\pi}{3}rh \\[3ex] \dfrac{\partial V}{\partial h} = \dfrac{\pi}{3}r^2 \end{array}\right\}$$

The variation when both the radius and the height change is given by $dV = \dfrac{2\pi}{3}rh\, dr + \dfrac{\pi}{3}r^2 dh$.

Example (4).   In the following example F and f denote two arbitrary functions of any form whatsoever. For example, they may be sine-functions, or exponentials, or mere algebraic functions of the two independent variables, t and x. This being understood, let us take the expression

$$y = F(x + at) + f(x - at)$$

or,

$$y = F(w) + f(v)$$

where

$$w = x + at, \quad \text{and} \quad v = x - at$$

Then

$$\frac{\partial y}{\partial x} = \frac{\partial F(w)}{\partial w} \cdot \frac{\partial w}{\partial x} + \frac{\partial f(v)}{\partial v} \cdot \frac{\partial v}{\partial x}$$

$$= F'(w) \cdot 1 + f'(v) \cdot 1$$

(where the figure 1 is simply the coefficient of x in w and v);

and

$$\frac{\partial^2 y}{\partial x^2} = F''(w) + f''(v)$$

Also

$$\frac{\partial y}{\partial t} = \frac{\partial F(w)}{\partial w} \cdot \frac{\partial w}{\partial t} + \frac{\partial f(v)}{\partial v} \cdot \frac{\partial v}{\partial t}$$

$$= F'(w) \cdot a - f'(v)a$$

and

$$\frac{\partial^2 y}{\partial t^2} = F''(w)a^2 + f''(v)a^2$$

whence

$$\frac{\partial^2 y}{\partial t^2} = a^2 \frac{\partial^2 y}{\partial x^2}$$

This differential equation is of immense importance in mathematical physics.

Maxima and Minima of Functions of Two Independent Variables

Example (5).   Let us take up again Exercises IX, No. 4.

Let x and y be the lengths of two of the portions of the string. The third is $30 - (x + y)$, and the area of the triangle is $A = \sqrt{s(s - x)(s - y)(s - 30 + x + y)}$, where s is the half perimeter, so that $s = 15$, and $A = \sqrt{15P}$, where

$$P = (15 - x)(15 - y)(x + y - 15)$$
$$= xy^2 + x^2y - 15x^2 - 15y^2 - 45xy + 450x + 450y - 3375$$

Clearly A is a maximum when P is maximum.

$$dP = \frac{\partial P}{\partial x}dx + \frac{\partial P}{\partial y}dy$$

For a maximum (clearly it will not be a minimum in this case), one must have simultaneously

$$\frac{\partial P}{\partial x} = 0 \quad \text{and} \quad \frac{\partial P}{\partial y} = 0$$

that is,
$$\left. \begin{array}{l} 2xy - 30x + y^2 - 45y + 450 = 0, \\ 2xy - 30y + x^2 - 45x + 450 = 0. \end{array} \right\}$$

Subtracting the second equation from the first and factoring gives

$$(y - x)(x + y - 15) = 0$$

so either $x = y$ or $x + y - 15 = 0$. In the latter case $P = 0$, which is not a maximum, hence $x = y$.

If we now introduce this condition in the value of P, we find

$$P = (15 - x)^2(2x - 15) = 2x^3 - 75x^2 + 900x - 3375$$

For maximum or minimum, $\dfrac{dP}{dx} = 6x^2 - 150x + 900 = 0$, which gives $x = 15$ or $x = 10$.

Clearly $x = 15$ gives zero area; $x = 10$ gives the maximum, for

$\dfrac{d^2P}{dx^2} = 12x - 150$, which is $+30$ for $x = 15$ and -30 for $x = 10$.

Example (6).   Find the dimensions of an ordinary railway coal truck with rectangular ends, so that, for a given volume V, the area of sides and floor together is as small as possible.

The truck is a rectangular box open at the top. Let x be the length and y be the width; then the depth is $\dfrac{V}{xy}$. The surface area

is $S = xy + \dfrac{2V}{x} + \dfrac{2V}{y}$.

$$dS = \frac{\partial S}{\partial x}dx + \frac{\partial S}{\partial y}dy = \left(y - \frac{2V}{x^2}\right)dx + \left(x - \frac{2V}{y^2}\right)dy$$

For minimum (clearly it won't be a maximum here),

$$y - \frac{2V}{x^2} = 0 \quad x - \frac{2V}{y^2} = 0$$

Multiplying the first equation by x, the second by y, and subtracting gives $x = y$. So $x^3 = 2V$ and $x = y = \sqrt[3]{2V}$.

EXERCISES XV

(1) Differentiate the expression $\dfrac{x^3}{3} - 2x^3y - 2y^2x + \dfrac{y}{3}$ with respect to x alone, and with respect to y alone.

(2) Find the partial derivatives with respect to x, y, and z, of the expression

$$x^2yz + xy^2z + xyz^2 + x^2y^2z^2$$

(3) Let $r^2 = (x - a)^2 + (y - b)^2 + (z - c)^2$.

Find the value of $\dfrac{\partial r}{\partial x} + \dfrac{\partial r}{\partial y} + \dfrac{\partial r}{\partial z}$. Also find the value of $\dfrac{\partial^2 r}{\partial x^2} + \dfrac{\partial^2 r}{\partial y^2} + \dfrac{\partial^2 r}{\partial z^2}$.

(4) Find the total derivative of $y = u^v$.

(5) Find the total derivative of $y = u^3 \sin v$; of $y = (\sin x)^u$; and of $y = \dfrac{\ln u}{v}$.

(6) Verify that the sum of three quantities x, y, z, whose product is a constant k, is minimum when these three quantities are equal.

(7) Does the function $u = x + 2xy + y$ have a maximum or minimum?

(8) A post office regulation once stated that no parcel is to be of such a size that its length plus its girth exceeds 6 feet. What is the greatest volume that can be sent by post (*a*) in the case of a package of rectangular cross-section; (*b*) in the case of a package of circular cross-section?

(9) Divide π into 3 parts such that the product of their sines may be a maximum or minimum.

(10) Find the maximum or minimum of $u = \dfrac{e^{x+y}}{xy}$.

(11) Find maximum and minimum of

$$u = y + 2x - 2 \ln y - \ln x$$

(12) A bucket of given capacity has the shape of a horizontal isosceles triangular prism with the apex underneath, and the opposite face open. Find its dimensions in order that the least amount of iron sheet may be used in its construction.

Chapter XVII

INTEGRATION

The great secret has already been revealed that this mysterious symbol $\int$, which is after all only a long S, merely means "the sum of", or "the sum of all such quantities as". It therefore resembles that other symbol Σ (the Greek *Sigma*), which is also a sign of summation. There is this difference, however, in the practice of mathematical men as to the use of these signs, that while Σ is generally used to indicate the sum of a number of finite quantities, the integral sign $\int$ is generally used to indicate the summing up of a vast number of small quantities of infinitely minute magnitude, mere elements in fact, that go to make up the total required. Thus $\int dy = y$, and $\int dx = x$.

Any one can understand how the whole of anything can be conceived of as made up of a lot of little bits; and the smaller the bits the more of them there will be. Thus, a line one inch long may be conceived as made up of 10 pieces, each $\frac{1}{10}$ of an inch long; or of 100 parts, each part being $\frac{1}{100}$ of an inch long; or of 1,000,000 parts, each of which is $\frac{1}{1,000,000}$ of an inch long; or, pushing the thought to the limits of conceivability, it may be regarded as made up of an infinite number of elements each of which is infinitesimally small.

Yes, you will say, but what is the use of thinking of anything that way? Why not think of it straight off, as a whole? The simple reason is that there are a vast number of cases in which one cannot calculate the bigness of the thing as a whole without reckoning up the sum of a lot of small parts. The process of *"integrating"* is to enable us to calculate totals that otherwise we should be unable to estimate directly.

Let us first take one or two simple cases to familiarize ourselves with this notion of summing up a lot of separate parts.

Consider the series:

$$1 + \tfrac{1}{2} + \tfrac{1}{4} + \tfrac{1}{8} + \tfrac{1}{16} + \tfrac{1}{32} + \tfrac{1}{64} + \ldots.$$

Here each member of the series is formed by taking half the value of the preceding one. What is the value of the total if we could go on to an infinite number of terms? Every schoolboy knows that the answer is 2. Think of it, if you like, as a line. Begin with one inch; add a half inch; add a quarter; add an eighth; and so on. If at any point of the operation we stop, there will still be a piece wanting to make up the whole 2 inches; and the piece wanting will always be the same size as the last piece added. Thus, if after having put together 1, $\tfrac{1}{2}$, and $\tfrac{1}{4}$, we stop, there will be $\tfrac{1}{4}$ wanting. If we go on till we have added $\tfrac{1}{64}$, there will still be $\tfrac{1}{64}$ wanting. The remainder needed will always be equal to the last term added. By an infinite number of operations only should we reach the actual 2 inches. Practically we should reach it when we got to pieces so small that they could not be drawn—that would be after about 10 terms, for the eleventh term is $\tfrac{1}{1024}$. If we want to go so far that no measuring machine could detect it, we should merely have to go to about 20 terms. A microscope would not show even the 18th term! So the infinite number of operations is no such dreadful thing after all. The *integral* is simply the whole lot. But, as we shall see, there are cases in which the integral calculus enables us to get at the *exact* total that there would be as the result of an infinite number of operations. In such cases the integral calculus gives us a *rapid* and easy way of getting at a result

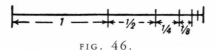

FIG. 46.

that would otherwise require an interminable lot of elaborate working out. So we had best lose no time in learning *how to integrate.*

Slopes of Curves, and the Curves Themselves

Let us make a little preliminary enquiry about the slopes of curves. For we have seen that differentiating a curve means finding an expression for its slope (or for its slopes at different points). Can we perform the reverse process of reconstructing the whole curve if the slope (or slopes) are prescribed for us?

Go back to Chapter 10, case (2). Here we have the simplest of curves, a sloping line with the equation

$$y = ax + b$$

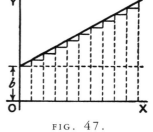

FIG. 47.

We know that here b represents the initial height of y when $x = 0$, and that a, which is the same as $\dfrac{dy}{dx}$, is the "slope" of the line. The line has a constant slope. All along it the elementary triangles ◿ $\mathbf{dy}$ have the same proportion between height and $\mathbf{dx}$ base. Suppose we were to take the dx's and dy's of finite magnitude, so that 10 dx's made up one inch, then there would be ten little triangles like

◿◿◿◿◿◿◿◿◿◿

Now, suppose that we were to reconstruct the "curve", starting merely from the information that $\dfrac{dy}{dx} = a$. What could we do? Still taking the little d's as of finite size, we could draw 10 of them, all with the same slope, and then put them together, end to end, like this:

And, as the slope is the same for all, they would join to make, as in Fig. 48, a sloping line sloping with the correct slope

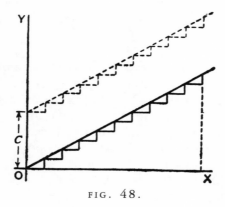

FIG. 48.

$\dfrac{dy}{dx} = a$. And whether we take the dy's and dx's as finite or infi-

nitely small, as they are all alike, clearly $\dfrac{y}{x} = a$, if we reckon y as

the total of all the dy's, and x as the total of all the dx's. But whereabouts are we to put this sloping line? Are we to start at the origin O, or higher up? As the only information we have is as to the slope, we are without any instructions as to the particular height above O; in fact the initial height is undetermined. The slope will be the same, whatever the initial height. Let us therefore make a shot at what may be wanted, and start the sloping line at a height C above O. That is, we have the equation

$$y = ax + C$$

It becomes evident now that in this case the added constant means the particular value that y has when $x = 0$.

Now let us take a harder case, that of a line, the slope of which is not constant, but turns up more and more. Let us assume that the upward slope gets greater and greater in proportion as x grows. In symbols this is:

$$\frac{dy}{dx} = ax$$

Or, to give a concrete case, take $a = \frac{1}{5}$, so that

$$\frac{dy}{dx} = \frac{1}{5}x$$

Then we had best begin by calculating a few of the values of the slope at different values of x, and also draw little diagrams of them.

When

$$x = 0, \quad \frac{dy}{dx} = 0,$$

$$x = 1, \quad \frac{dy}{dx} = 0 \cdot 2,$$

$$x = 2, \quad \frac{dy}{dx} = 0 \cdot 4,$$

$$x = 3, \quad \frac{dy}{dx} = 0 \cdot 6,$$

$$x = 4, \quad \frac{dy}{dx} = 0 \cdot 8,$$

$$x = 5, \quad \frac{dy}{dx} = 1 \cdot 0.$$

Now try to put the pieces together, setting each so that the middle of its base is the proper distance to the right, and so that they fit together at the corners; thus (Fig. 49). The result is, of course, not a smooth curve: but it is an approximation to one. If we had taken bits half as long, and twice as numerous, like Fig. 50, we should have a better approximation.[1]

1. Approximating a continuous curve by drawing smaller and smaller right triangles under the curve is today called the "trapezoidal rule" because the little triangles, joined to the narrow rectangles beneath them, form trapezoids as shown in Figure 47.

A closer approximation, though much more difficult to apply, is to draw tiny parabolas below (or above) a curve's segments. The sum is then approximated by the "parabolic rule" or what is also known as "Simpson's rule" after British mathematician Thomas Simpson (1710-1761). Thompson does not go into this, but you can read about Simpson's rule in modern calculus textbooks.—M.G.

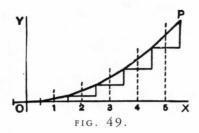

FIG. 49.

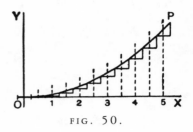

FIG. 50.

But for a perfect curve we ought to take each dx and its corresponding dy infinitesimally small, and infinitely numerous.

Then, how much ought the value of any y to be? Clearly, at any point P of the curve, the value of y will be the sum of all the little dy's from 0 up to that level, that is to say, $\int dy = y$. And as each dy is equal to $\frac{1}{5}x \cdot dx$, it follows that the whole y will be equal to the sum of all such bits as $\frac{1}{5}x \cdot dx$, or, as we should write it,

$$\int \tfrac{1}{5}x \cdot dx.$$

Now if x had been constant, $\int \frac{1}{5}x \cdot dx$ would have been the same as $\frac{1}{5}x \int dx$, or $\frac{1}{5}x^2$. But x began by being 0, and increases to the particular value of x at the point P, so that its average value from 0 to that point is $\frac{1}{2}x$. Hence $\int \frac{1}{5}x\,dx = \frac{1}{10}x^2$; or $y = \frac{1}{10}x^2$.

But, as in the previous case, this requires the addition of an undetermined constant C, because we have not been told at what height above the origin the curve will begin, when $x = 0$. So we write, as the equation of the curve drawn in Fig. 51,

$$y = \tfrac{1}{10}x^2 + C$$

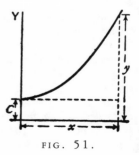

FIG. 51.

EXERCISES XVI

(1) Find the ultimate sum of $\frac{2}{3} + \frac{1}{3} + \frac{1}{6} + \frac{1}{12} + \frac{1}{24} + \ldots$

(2) Show that the series $1 - \frac{1}{2} + \frac{1}{3} - \frac{1}{4} + \frac{1}{5} - \frac{1}{6} + \frac{1}{7} \ldots$, is convergent, and find its sum to 8 terms.

(3) If $\ln (1 + x) = x - \dfrac{x^2}{2} + \dfrac{x^3}{3} - \dfrac{x^4}{4} + \ldots$, find $\ln 1.3$.

(4) Following a reasoning similar to that explained in this chapter, find y,

$$\text{if } \frac{dy}{dx} = \tfrac{1}{4}x$$

(5) If $\dfrac{dy}{dx} = 2x + 3$, find y

Chapter XVIII

INTEGRATING AS THE REVERSE OF DIFFERENTIATING

━━━━ ∞∞∞ ━━━━

Differentiating is the process by which when y is given us as a function of x, we can find $\dfrac{dy}{dx}$.

Like every other mathematical operation, the process of differentiation may be reversed.[1] Thus, if differentiating $y = x^4$ gives us $\dfrac{dy}{dx} = 4x^3$, then, if one begins with $\dfrac{dy}{dx} = 4x^3$, one would say that reversing the process would yield $y = x^4$. But here comes in a curious point. We should get $\dfrac{dy}{dx} = 4x^3$ if we had begun with *any* of the following: x^4, or $x^4 + a$, or $x^4 + c$, or x^4 with *any* added constant. So it is clear that in working backwards from $\dfrac{dy}{dx}$ to y,

---

1. Familiar examples of inverse operations in arithmetic are subtraction as the reverse of addition, division as the reverse of multiplication, and root extraction as the reverse of raising to higher powers.

In arithmetic you can test a subtraction $A - B = C$ by adding B and C to see if you get A. In similar fashion one can test an integration or a differentiation by reversing the process to see if you return to the original expression.

In its geometrical model, differentiating a function gives a formula that determines the slope of a function's curve at any given point. Integration provides a method with which, given the formula for the slope you can determine the curve and its function. This in turn provides a quick way of calculating the area between intervals on a curve and the graph's x axis.—M.G.

one must make provision for the possibility of there being an added constant, the value of which will be undetermined until ascertained in some other way. So, if differentiating x^n yields nx^{n-1}, going backwards from $\dfrac{dy}{dx} = nx^{n-1}$ will give us $y = x^n + C$;

where C stands for the yet undetermined possible constant.

Clearly, in dealing with powers of x, the rule for working backwards will be: Increase the power by 1, then divide by that increased power, and add the undetermined constant.

So, in the case where

$$\frac{dy}{dx} = x^n$$

working backwards, we get

$$y = \frac{1}{n+1}x^{n+1} + C$$

If differentiating the equation $y = ax^n$ gives us

$$\frac{dy}{dx} = anx^{n-1}$$

it is a matter of common sense that beginning with

$$\frac{dy}{dx} = anx^{n-1}$$

and reversing the process, will give us

$$y = ax^n$$

So, when we are dealing with a multiplying constant, we must simply put the constant as a multiplier of the result of the integration.

Thus, if $\dfrac{dy}{dx} = 4x^2$, the reverse process gives us $y = \tfrac{4}{3}x^3$.

But this is incomplete. For we must remember that if we had started with

$$y = ax^n + C$$

where C is any constant quantity whatever, we should equally have found

$$\frac{dy}{dx} = anx^{n-1}$$

So, therefore, when we reverse the process we must always remember to add on this undetermined constant, even if we do not yet know what its value will be.[2]

This process, the reverse of differentiating, is called *integrating*; for it consists in finding the value of the whole quantity y when you are given only an expression for dy or for $\frac{dy}{dx}$. Hitherto we have as much as possible kept dy and dx together as a derivative: henceforth we shall more often have to separate them.

If we begin with a simple case,

$$\frac{dy}{dx} = x^2$$

We may write this, if we like, as

$$dy = x^2 dx$$

2. A joke about integration recently made the rounds of math students and teachers. Two students at a technical college, Bill and Joe, are having lunch at a campus hangout. Bill complains that math teaching has become so poor in the United States that most college students know next to nothing about calculus.

Joe disagrees. While Bill is in the men's room, Joe calls over a pretty blond waitress. He gives her five dollars to play a joke on Bill. When she brings the dessert he will ask her a question. He doesn't tell her the question, but he instructs her to answer "one third x cubed." The waitress smiles, pockets the fiver, and agrees.

When Bill returns to the booth, Joe proposes the following twenty-dollar bet. He will ask their waitress about an integral. If she responds correctly he wins the bet. Joe knows he can't lose. The two friends shake hands on the deal.

When the waitress comes to the table, Joe asks, "What's the integral of x squared?"

"One third x cubed," she replies. Then as she walks away she says over her shoulder, "plus a constant."—M.G.

Now this is a "differential equation" which informs us that an element of y is equal to the corresponding element of x multiplied by x^2. Now, what we want is the integral; therefore, write down with the proper symbol the instructions to integrate both sides, thus:

$$\int dy = \int x^2 dx$$

[Note as to reading integrals: the above would be read thus:

"Integral of dee-wy equals *integral of eks-squared dee-eks."*]

We haven't yet integrated: we have only written down instructions to integrate—if we can. Let us try. Plenty of other fools can do it—why not we also? The left-hand side is simplicity itself. The sum of all the bits of y is the same thing as y itself. So we may at once put:

$$y = \int x^2 dx$$

But when we come to the right-hand side of the equation we must remember that what we have got to sum up together is not all the dx's, but all such terms as $x^2 dx$; and this will *not* be the same as $x^2 \int dx$, because x^2 is not a constant. For some of the dx's will be multiplied by big values of x^2, and some will be multiplied by small values of x^2, according to what x happens to be. So we must bethink ourselves as to what we know about this process of integration being the reverse of differentiation. Now, our rule for this reversed process when dealing with x^n is "increase the power by one, and divide by the same number as this increased power". That is to say, $x^2 dx$ will be changed* to $\frac{1}{3}x^3$. Put this into

*You may ask: what has become of the little dx at the end? Well, remember that it was really part of the derivative, and when changed over to the right-hand side, as in the $x^2 dx$, serves as a reminder that x is the independent variable with respect to which the operation is to be effected; and, as the result of the product being totalled up, the power of x has increased by *one.* You will soon become familiar with all this.

the equation; but don't forget to add the "constant of integration" C at the end. So we get:

$$y = \tfrac{1}{3}x^3 + C$$

You have actually performed the integration. How easy!
Let us try another simple case

Let
$$\frac{dy}{dx} = ax^{12}$$

where a is any constant multiplier. Well, we found when differentiating (see Chapter V) that any constant factor in the value of y reappeared unchanged in the value of $\dfrac{dy}{dx}$. In the reversed process of integrating, it will therefore also reappear in the value of y. So we may go to work as before, thus:

$$dy = ax^{12} \cdot dx$$

$$\int dy = \int ax^{12} \cdot dx$$

$$\int dy = a \int x^{12} dx$$

$$y = a \times \tfrac{1}{13}x^{13} + C$$

So that is done. How easy!

We begin to realize now that integrating is a process *of finding our way back,* as compared with differentiating. If ever, during differentiating, we have found any particular expression—in this example ax^{12}—we can find our way back to the y from which it was derived. The contrast between the two processes may be illustrated by the following illustration due to a well-known teacher. If a stranger to Manhattan were set down in Times Square, and told to find his way to Grand Central Station, he might find the task hopeless. But if he had previously been personally conducted from Grand Central Station to Times Square, it would be comparatively easy for him to find his way back to Grand Central Station.

Integration of the Sum or Difference of Two Functions

Let
$$\frac{dy}{dx} = x^2 + x^3$$

then
$$dy = x^2 dx + x^3 dx$$

There is no reason why we should not integrate each term separately; for, as may be seen in Chapter VI, we found that when we differentiated the sum of two separate functions, the derivative was simply the sum of the two separate differentiations. So, when we work backwards, integrating, the integration will be simply the sum of the two separate integrations.

Our instructions will then be:

$$\int dy = \int (x^2 + x^3) dx$$

$$= \int x^2 dx + \int x^3 dx$$

$$y = \tfrac{1}{3}x^3 + \tfrac{1}{4}x^4 + C$$

If either of the terms had been a negative quantity, the corresponding term in the integral would have also been negative. So that differences are as readily dealt with as sums.

How to Deal with Constant Terms

Suppose there is in the expression to be integrated a constant term—such as this:

$$\frac{dy}{dx} = x^n + b$$

This is laughably easy. For you have only to remember that when you differentiated the expression $y = ax$, the result was $\frac{dy}{dx} = a$. Hence, when you work the other way and integrate, the constant reappears multiplied by x. So we get

$$dy = x^n dx + b \cdot dx$$

$$\int dy = \int x^n dx + \int b \, dx$$

$$y = \frac{1}{n+1} x^{n+1} + bx + C$$

Here are a lot of examples on which to try your newly acquired powers.

Examples.

(1) Given $\dfrac{dy}{dx} = 24x^{11}$. Find y. *Ans.* $y = 2x^{12} + C$

(2) Find $\displaystyle\int (a+b)(x+1)dx$. It is $(a+b)\displaystyle\int (x+1)dx$

or　　　$(a+b)\left[\displaystyle\int x \, dx + \displaystyle\int dx\right]$ or $(a+b)\left(\dfrac{x^2}{2} + x\right) + C$

(3) Given $\dfrac{du}{dt} = gt^{\frac{1}{2}}$ Find u. *Ans.* $u = \frac{2}{3}gt^{\frac{3}{2}} + C$

(4) $\dfrac{dy}{dx} = x^3 - x^2 + x$ Find y.

$$dy = (x^3 - x^2 + x)\,dx$$

or　　$dy = x^3 dx - x^2 dx + x \, dx;\ y = \displaystyle\int x^b dx - \displaystyle\int x^2 dx + \displaystyle\int x \, dx$

and　　$y = \frac{1}{4}x^4 - \frac{1}{3}x^3 + \frac{1}{2}x^2 + C$

(5) Integrate $9.75x^{2.25}dx$. *Ans.* $y = 3x^{3.25} + C$

All these are easy enough. Let us try another case.

Let　　　　　　　　　　　$\dfrac{dy}{dx} = ax^{-1}$

Proceeding as before, we will write

$$dy = ax^{-1} \cdot dx, \quad \int dy = a \int x^{-1} dx$$

Well, but what is the integral of $x^{-1} dx$?

If you look back amongst the results of differentiating x^2 and x^3 and x^n, etc., you will find we never got x^{-1} from any one of them as the value of $\dfrac{dy}{dx}$. We got $3x^2$ from x^3; we got $2x$ from x^2;

we got 1 from x^1 (that is, from x itself); but we did not get x^{-1} from x^0, for a very good reason. Its derivative (got by slavishly following the usual rule) is $0 \times x^{-1}$, and that multiplication by zero gives it zero value! Therefore when we now come to try to integrate $x^{-1} dx$, we see that it does not come in anywhere in the powers of x that are given by the rule:

$$\int x^n dx = \frac{1}{n+1} x^{n+1}$$

It is an exceptional case.

Well; but try again. Look through all the various derivatives obtained from various functions of x, and try to find amongst them x^{-1}. A sufficient search will show that we actually did get

$\dfrac{dy}{dx} = x^{-1}$ as the result of differentiating the function $y = \ln x$.

Then, of course, since we know that differentiating $\ln x$ gives us x^{-1}, we know that, by reversing the process, integrating $dy = x^{-1} dx$ will give us $y = \ln x$. But we must not forget the constant factor a that was given, nor must we omit to add the undetermined constant of integration. This then gives us as the solution to the present problem,

$$y = a \ln x + C$$

But this is valid only for $x > 0$. For $x < 0$, you should verify that the solution is

$$y = a \ln (-x) + C$$

These two cases are usually combined by writing

$$y = a \ln |x| + C$$

where $|x|$ is the absolute value of x, namely x if $x \geq 0$ and $-x$ if $x < 0$.

N.B.—Here note this very remarkable fact, that we could not have integrated in the above case if we had not happened to know the corresponding differentiation. If no one had found out that differentiating $\ln x$ gave x^{-1}, we should have been utterly stuck by the problem how to integrate $x^{-1}dx$. Indeed it should be frankly admitted that this is one of the curious features of the integral calculus:—that you can't integrate anything before the reverse process of differentiating something else has yielded that expression which you want to integrate.

Another Simple Case.

Find $\int (x + 1)(x + 2)dx$.

On looking at the function to be integrated, you remark that it is the product of two different functions of x. You could, you think, integrate $(x + 1)dx$ by itself, or $(x + 2)dx$ by itself. Of course you could. But what to do with a product? None of the differentiations you have learned have yielded you for the derivative a product like this. Failing such, the simplest thing is to multiply up the two functions, and then integrate. This gives us

$$\int (x^2 + 3x + 2)dx$$

And this is the same as

$$\int x^2dx + \int 3x\, dx + \int 2\, dx$$

And performing the integrations, we get

$$\tfrac{1}{3}x^3 + \tfrac{3}{2}x^2 + 2x + C$$

Some Other Integrals

Now that we know that integration is the reverse of differentiation, we may at once look up the derivatives we already know, and see from what functions they were derived. This gives us the following integrals ready made:

$$x^{-1} \qquad \int x^{-1} dx \qquad = \ln |x| + C$$

$$\frac{1}{x+a} \qquad \int \frac{1}{x+a} dx \qquad = \ln |x+a| + C$$

$$e^x \qquad \int e^x dx \qquad = e^x + C$$

$$e^{-x} \qquad \int e^{-x} dx \qquad = -e^{-x} + C$$

for if $y = \dfrac{-1}{e^x}$, $\dfrac{dy}{dx} = -\dfrac{e^x \times 0 - 1 \times e^x}{e^{2x}} = e^{-x}$

$$\sin x; \qquad \int \sin x \, dx \qquad = -\cos x + C$$

$$\cos x; \qquad \int \cos x \, dx \qquad = \sin x + C$$

Also we may deduce the following:

$$\ln x \qquad \int \ln x \, dx \qquad = x(\ln x - 1) + C$$

for if $y = x \ln x - x$, $\dfrac{dy}{dx} = \dfrac{x}{x} + \ln x - 1 = \ln x$

$$\log_{10} x \int \log_{10} \times dx = 0.4343x(\ln x - 1) + C$$

$$a^x \qquad \int a^x dx \qquad = \frac{a^x}{\ln a} + C$$

$$\cos ax \int \cos ax \ dx = \frac{1}{a} \sin ax + C$$

for if $y = \sin ax$, $\frac{dy}{dx} = a \cos ax$; hence to get $\cos ax$ one must differentiate $y = \frac{1}{a} \sin ax$

$$\sin ax \int \sin ax \ dx = -\frac{1}{a} \cos ax + C$$

Try also $\cos^2 \theta$; a little dodge will simplify matters:

$$\cos 2\theta = \cos^2 \theta - \sin^2 \theta = 2 \cos^2 \theta - 1$$

hence $\qquad\qquad \cos^2 \theta = \frac{1}{2}(\cos 2\theta + 1)$

and $$\int \cos^2 \theta \ d\theta = \frac{1}{2} \int (\cos 2\theta + 1) d\theta$$

$$= \frac{1}{2} \int \cos 2\theta \ d\theta + \frac{1}{2} \int d\theta$$

$$= \frac{\sin 2\theta}{4} + \frac{\theta}{2} + C.$$

See also the Table of Standard Forms at the end of the last chapter. You should make such a table for yourself, putting in it only the general functions which you have successfully differentiated and integrated. See to it that it grows steadily!

EXERCISES XVII

(1) Find $\displaystyle\int y\,dx$ when $y^2 = 4ax$.

(2) Find $\displaystyle\int \frac{3}{x^4}\,dx$.

(3) Find $\displaystyle\int \frac{1}{a}\,x^3\,dx$.

(4) Find $\displaystyle\int (x^2 + a)dx$.

(5) Integrate $5x^{-\frac{7}{2}}$.

(6) Find $\displaystyle\int (4x^3 + 3x^2 + 2x + 1)dx$.

(7) If $\dfrac{dy}{dx} = \dfrac{ax}{2} + \dfrac{bx^2}{3} + \dfrac{cx^3}{4}$; find y.

(8) Find $\displaystyle\int \left(\frac{x^2 + a}{x + a} \right) dx$.

(9) Find $\displaystyle\int (x + 3)^3 dx$.

(10) Find $\displaystyle\int (x + 2)(x - a)dx$.

(11) Find $\displaystyle\int \left(\sqrt{x} + \sqrt[3]{x} \right) 3a^2\,dx$.

(12) Find $\displaystyle\int \left(\sin \theta - \tfrac{1}{2} \right) \frac{d\theta}{3}$.

(13) Find $\displaystyle\int \cos^2 a\theta\,d\theta$.

(14) Find $\displaystyle\int \sin^2 \theta\,d\theta$.

(15) Find $\displaystyle\int \sin^2 a\theta\,d\theta$.

(16) Find $\displaystyle\int e^{3x}\,dx$.

(17) Find $\displaystyle\int \frac{dx}{1 + x}$.

(18) Find $\displaystyle\int \frac{dx}{1 - x}$.

Chapter XIX

ON FINDING AREAS
BY INTEGRATING

———⟨∞⟩———

One use of the integral calculus is to enable us to ascertain the values of areas bounded by curves.

Let us try to get at the subject bit by bit.

Let AB be a curve, the equation to which is known. That is, y in this curve is some known function of x. (See Fig. 52.) Think of a piece of the curve from the point P to the point Q.

Let a perpendicular PM be dropped from P, and another QN from the point Q. Then call $OM = x_1$ and $ON = x_2$, and the ordinates $PM = y_1$ and $QN = y_2$. We have thus marked out the area $PQNM$ that lies beneath the piece PQ. The problem is, *how can we calculate the value of this area?*

The secret of solving this problem is to conceive the area as being divided up into a lot of narrow strips, each of them being of the width dx.[1] The smaller we take dx, the more of them there will be between x_1 and x_2. Now, the whole area is clearly equal to the sum of the areas of all such strips. Our business will then

1. Thompson explains the integral as the sum of a finite number of thin strips under the curve as their widths approach the limit of zero. When these strips are all below a curve their sum is called today a "lower Riemann sum" after German mathematician George Friedrich Bernhard Riemann (1826–1866). The same sum can be obtained by letting the strips extend above the curve as shown in Figure 42, in which case the sum is called an "upper Riemann sum." If the strips are drawn so that their tops cross the curve, their sum is called a "Riemann sum." Regardless of how the strips are drawn, they will have the same "Riemann integral" at the limit when they became infinite in number and their widths become infinitely small.—M.G.

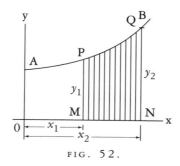

be to discover an expression for the area of any one narrow strip, and to integrate it so as to add together all the strips. Now think of any one of the strips. It will be like this: being bounded between two vertical sides, with a flat bottom dx, and with a slightly curved sloping top. Suppose we take its *average* height as being y; then, as its width is dx, its area will be $y dx$. And seeing that we may take the width as narrow as we please, if we only take it narrow enough its average height will be the same as the height at the middle of it. Now let us call the unknown value of the whole area S, meaning surface. The area of one strip will be simply a bit of the whole area, and may therefore be called dS. So we may write

$$\text{area of 1 strip} = dS = y \, dx$$

If then we add up all the strips, we get

$$\text{total area } S = \int dS = \int y \, dx$$

So then our finding S depends on whether we can integrate $y \, dx$ for the particular case, when we know what the value of y is as a function of x.

For instance, if you were told that for the particular curve in question $y = b + ax^2$, no doubt you could put that value into the expression and say: then I must find $\int (b + ax^2) dx$.

That is all very well; but a little thought will show you that something more must be done. Because the area we are trying to find is not the area under the whole length of the curve, but only the area limited on the left by *PM*, and on the right by *QN*, it follows that we must do something to define our area between those bounds.

This introduces us to a new notion, namely, that of *integrating between limits.*[2] We suppose x to vary, and for the present purpose we do not require any value of x below x_1 (that is *OM*), nor any value of x above x_2 (that is *ON*). When an integral is to be thus defined between two limits, we call the lower of the two values *the inferior limit,* and the upper value *the superior limit.* Any integral so limited we designate as a *definite integral,* by way of distinguishing it from a *general integral* to which no limits are assigned.[3]

2. The word "limit" is confusing here because it is not a limit in the sense of the sum of an infinite series. The word "bound" is much clearer. What Thompson calls inferior and superior limits of a closed interval along a continuous curve are lower and upper bounds, although today many textbooks call them "lower and upper limits of integration," or "left and right endpoints of integration."—M.G.

3. Thompson's term "general integral" is no longer used. In the past it was also called a "primitive integral" and later an "indefinite integral." Today it is usually called an "antiderivative." The reason is obvious. It is the inverse of a derivative. Writers differ on how to symbolize it. Thompson simply puts it inside brackets. One common symbol for it today is $F(x)$, using a capital F instead of a lower case f. In all that follows I have substituted "antiderivative" for Thompson's "general integral."

As Thompson makes clear, a derivative does not have a unique antiderivative because an antiderivative can have any one of an infinite number of added constants. These constants correspond to different heights a curve can have above the x axis. For example, the derivative of x^2 is $2x$. But $2x$ is also the derivative of $x^2 + 1$; $x^2 + 666$; $x^2 - $ pi, and so on. It can be x^2 plus or minus any real number. Because there is an infinity of real numbers, if a derivative has one antiderivative it will have an infinite number of them. They differ only in what are called their "constants of integration." The antiderivative is "indefinite" because it is not unique.

In *Calculus Made Easy,* and in all calculus textbooks, when "integral" is used without an adjective, it means the definite integral. It is the fundamental concept of integration.—M.G.

In the symbols which give instructions to integrate, the limits are marked by putting them at the top and bottom respectively of the sign of integration. Thus the instruction

$$\int_{x=x_1}^{x=x_2} y \cdot dx$$

will be read: find the integral of $y \cdot dx$ between the inferior limit x_1 and the superior limit x_2.

Sometimes the thing is written more simply

$$\int_{x_1}^{x_2} y \cdot dx$$

Well, but *how* do you find an integral between limits when you have got these instructions?

Look again at Fig. 52. Suppose we could find the area under the larger piece of curve from A to Q, that is from $x = 0$ to $x = x_2$, naming the area $AQNO$. Then, suppose we could find the area under the smaller piece from A to P, that is from $x = 0$ to $x = x_1$, namely, the area $APMO$. If then we were to subtract the smaller area from the larger, we should have left as a remainder the area $PQNM$, which is what we want. Here we have the clue as to what to do; the definite integral between the two limits is *the difference* between the antiderivative worked out for the superior limit and the antiderivative worked out for the lower limit.

Let us then go ahead. First, find the antiderivative thus:

$$\int y \, dx$$

and, as $y = b + ax^2$ is the equation to the curve (Fig. 52),

$$\int (b + ax^2) dx$$

is the antiderivative which we must find.

Doing the integration in question, we get

$$bx + \frac{a}{3} x^3 + C$$

and this will be the whole area from 0 up to any value of x that we may assign. When x is 0, this area is 0, so $C = 0$.

Therefore, the larger area up to the superior limit x_2 will be

$$bx_2 + \frac{a}{3} x_2{}^3$$

and the smaller area up to the inferior limit x_1 will be

$$bx_1 + \frac{a}{3} x_1{}^3$$

Now, subtract the smaller from the larger, and we get for the area S the value,

$$\text{area } S = b(x_2 - x_1) + \frac{a}{3}(x_2{}^3 - x_1{}^3)$$

This is the answer we wanted. Let us give some numerical values. Suppose $b = 10$, $a = 0.06$, and $x_2 = 8$ and $x_1 = 6$. Then the area S is equal to

$$10(8 - 6) + \frac{0.06}{3}(8^3 - 6^3)$$

$$= 20 + 0.02(512 - 216)$$

$$= 20 + 0.02 \times 296$$

$$= 20 + 5.92$$

$$= 25.92$$

Let us here put down a symbolic way of stating what we have ascertained about limits:

$$\int_{x=x_1}^{x=x_2} y \, dx = y_2 - y_1$$

where y_2 is the integrated value of $y\ dx$ corresponding to x_2, and y_1 that corresponding to x_1.

All integration between limits requires the difference between two values to be thus found. Also note that, in making the subtraction the added constant C has disappeared.[4]

4. Because the technique Thompson describes is at the heart of integral calculus, let me try to make it clearer.

To transform an antiderivative into a definite integral, bounds on the continuous curve must be specified. Each bound has a value for the curve's antiderivative. The definite integral is the difference between those two values. Simply subtract the value of the antiderivative at the left bound, where x is smaller, from the value of the antiderivative at the right bound, where x is larger. The result is the definite integral.

The definite integral is not a function. It is a number that is the limit sum of all the thin rectangles under the curve, between the curve's upper and lower bounds, as their widths approach zero and their number becomes infinite. The situation is analogous to cutting a piece of string. Suppose it is a foot long and you wish to obtain a 9-inch portion from the 3-inch mark to the 12-inch end. What do you do? You snip off the first three inches.

The fact that the definite integral is the difference between two values of the antiderivative is known as the "fundamental theorem of calculus." The theorem can be expressed in other ways, but this way is the simplest and most useful. It is an amazing theorem—one that unites differentiating with integrating. It works like sorcery, almost too good to be true!

Jerry P. King, in his *Art of Mathematics* (1992) likens the theorem to the cornerstone of an arch that holds together the two sides of calculus. Because there is one unified calculus, many mathematicians have recommended dropping the terms "differential calculus" and "integral calculus," replacing them with "calculus of derivatives" and "calculus of integrals."

It is impossible, King writes, to overestimate the importance of this arch. "Above the great arch and supported by it rests all of mathematical analysis and the significant parts of physics and the other sciences which calculus sustains and explains. Mathematics and science *stand* on calculus. . . ."

Newton was the first to construct the arch. Eric Temple Bell, in his chapter on Newton in *Men of Mathematics* (1937), calls the arch "surely one of the most astonishing things a mathematician ever discovered."

The mean value theorem (see the Postscript of Chapter X) defines a point p on a continuous function curve between bounds a and b. The point's y value is the mean value of the function. If you draw a horizontal line through this point, and drop vertical lines through a and b to the x axis, you form what is called the function's "mean value rectangle." (See Figure 53a). The area of this rectangle, shown shaded, is exactly equal to the area under the curve's interval from a to b.—M.G.

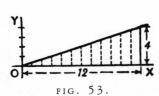

FIG. 53.

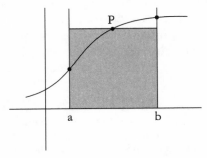

FIG. 53a. The mean-value rectangle.

Examples.

(1) To familiarize ourselves with the process, let us take a case of which we know the answer beforehand. Let us find the area of the triangle (Fig. 53), which has base $x = 12$ and height $y = 4$. We know beforehand, from obvious mensuration, that the answer will come to 24.

Now, here we have as the "curve" a sloping line for which the equation is

$$y = \frac{x}{3}$$

The area in question will be

$$\int_{x=0}^{x=12} y \cdot dx = \int_{x=0}^{x=12} \frac{x}{3} \cdot dx$$

Integrating $\frac{x}{3}\ dx$, and putting down the value of the anti-derivative in square brackets with the limits marked above and below, we get

$$\text{area} = \left[\frac{1}{3} \cdot \frac{1}{2}\, x^2 + C \right]_{x=0}^{x=12}$$

$$= \left[\frac{x^2}{6} + C \right]_{x=0}^{x=12}$$

$$\text{area} = \left[\frac{12^2}{6} + C \right] - \left[\frac{0^2}{6} + C \right]$$
$$= \frac{144}{6} = 24$$

Note that, in dealing with definite integrals, the constant C always disappears by subtraction.

Let us satisfy ourselves about this rather surprising dodge of calculation, by testing it on a simple example, Get some squared paper, preferably some that is ruled in little squares of one-eighth or one-tenth inch each way. On this squared paper plot out the graph of this equation,

$$y = \frac{x}{3}$$

The values to be plotted will be:

x	0	3	6	9	12
y	0	1	2	3	4

The plot is given in Fig. 54.

Now reckon out the area beneath the curve *by counting the little squares* below the line, from $x = 0$ as far as $x = 12$ on the right. There are 18 whole squares and four triangles, each of which has an area equal to $1\frac{1}{2}$ squares; or, in total, 24 squares. Hence 24 is

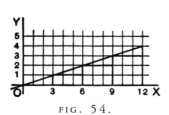

FIG. 54.

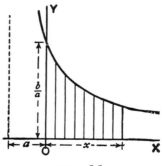

FIG. 55.

the numerical value of the integral of $\dfrac{x}{3}$ dx between the lower limit of $x = 0$ and the higher limit of $x = 12$.

As a further exercise, show that the value of the same integral between the limits of $x = 3$ and $x = 15$ is 36.

(2) Find the area, between limits $x = x_1$ and $x = 0$, of the curve

$$y = \frac{b}{x + a}$$

$$\text{Area} = \int_{x=0}^{x=x_1} y \cdot dx = \int_{x=0}^{x=x_1} \frac{b}{x + a}\, dx$$

$$= b \left[\ln (x + a) + C \right]_0^{x_1}$$

$$= b[\ln (x_1 + a) + C - \ln (0 + a) - C]$$

$$= b \ln \frac{x_1 + a}{a}$$

Let it be noted that this process of subtracting one part from a larger to find the difference is really a common practice. How do you find the area of a plane ring (Fig. 56), the outer radius of which is r_2 and the inner radius is r_1? You know from mensuration that the area of the outer circle is πr_2^2; then you find the area of the inner circle πr_1^2; then you subtract the latter from the former, and find area of ring $= \pi(r_2^2 - r_1^2)$; which may be written

$$\pi(r_2 + r_1)(r_2 - r_1)$$

$=$ mean circumference of ring $\times$ width of ring.

(3) Here's another case—that of *the die-away curve*. Find the area between $x = 0$ and $x = a$ of the curve (Fig. 57) whose equation is

$$y = be^{-x}$$

$$\text{Area} = b \int_{x=0}^{x=a} e^{-x} \cdot dx$$

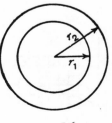

FIG. 56.

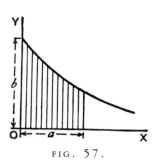

FIG. 57.

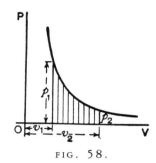

FIG. 58.

The integration gives

$$= b\left[-e^{-x}\right]_{0}^{a}$$

$$= b[-e^{-a} - (-e^{-0})]$$

$$= b(1 - e^{-a})$$

(4) Another example is afforded by the adiabatic curve of a perfect gas, the equation to which is $pv^{n} = c$, where p stands for pressure, v for volume, and n has the value 1.42 (Fig. 58).

Find the area under the curve (which is proportional to the work done in suddenly compressing the gas) from volume v_2 to volume v_1.

Here we have

$$\text{area} = \int_{v=v_1}^{v=v_2} cv^{-n} \cdot dv$$

$$= c\left[\frac{1}{1-n} v^{1-n}\right]_{v_1}^{v_2}$$

$$= c\,\frac{1}{1-n}(v_2^{1-n} - v_1^{1-n})$$

$$= \frac{-c}{0.42}\left(\frac{1}{v_2^{0.42}} - \frac{1}{v_1^{0.42}}\right)$$

An Exercise.

Prove the ordinary mensuration formula, that the area A of a circle whose radius is R, is equal to πR^2.

Consider an elementary zone or annulus of the surface (Fig. 59), of breadth dr, situated at a distance r from the centre. We may consider the entire surface as consisting of such narrow zones, and the whole area A will simply be the integral of all such elementary zones from centre to margin, that is, integrated from $r = 0$ to $r = R$.

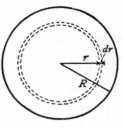

FIG. 59.

We have therefore to find an expression for the elementary area dA of the narrow zone. Think of it as a strip of breadth dr, and of a length that is the periphery of the circle of radius r, that is, a length of $2\pi r$. Then we have, as the area of the narrow zone,

$$dA = 2\pi r\, dr$$

Hence the area of the whole circle will be:

$$A = \int dA = \int_{r=0}^{r=R} 2\pi r \cdot dr = 2\pi \int_{r=0}^{r=R} r \cdot dr$$

Now, the antiderivative of $r \cdot dr$ is $\frac{1}{2}r^2$. Therefore,

$$A = 2\pi \left[\tfrac{1}{2}r^2\right]_{r=0}^{r=R}$$

or $A = 2\pi\left[\tfrac{1}{2}R^2 - \tfrac{1}{2}(0)^2\right]$

whence $A = \pi R^2$

Another Exercise.

Let us find the mean value of the positive part of the curve $y = x - x^2$, which is shown in Fig. 60. To find the mean ordinate, we shall have to find the area of the piece OMN, and then divide it by the length of the base ON. But before we can find the area we must ascertain the length of the base, so as

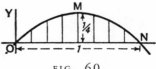

FIG. 60.

to know up to what limit we are to integrate. At N the ordinate y has zero value; therefore, we must look at the equation and see what value of x will make $y = 0$. Now, clearly, if x is 0, y will also be 0, the curve passing through the origin O; but also, if $x = 1$, $y = 0$: so that $x = 1$ gives us the position of the point N.

Then the area wanted is

$$= \int_{x=0}^{x=1} (x - x^2)dx = \left[\tfrac{1}{2}x^2 - \tfrac{1}{3}x^3\right]_0^1 = \left[\tfrac{1}{2} - \tfrac{1}{3}\right] - [0 - 0] = \tfrac{1}{6}$$

But the base length is 1.

Therefore, the average ordinate of the curve $= \tfrac{1}{6}$.

[*N.B.*—It will be a pretty and simple exercise in maxima and minima to find by differentiation what is the height of the maximum ordinate. It *must* be greater than the average.]

The mean ordinate of any curve, over a range from $x = 0$ to $x = x_1$, is given by the expression,

$$\text{mean } y = \frac{1}{x_1} \int_{x=0}^{x=x_1} y \cdot dx$$

If the mean ordinate be required over a distance not beginning at the origin but beginning at a point distant x_1 from the origin and ending at a point distant x_2 from the origin, the value will be

$$\text{mean } y = \frac{1}{x_2 - x_1} \int_{x=x_1}^{x=x_2} y \, dx$$

Areas in Polar Coordinates

When the equation of the boundary of an area is given as a function of the distance r of a point of it from a fixed point O (see Fig. 61) called the *pole,* and of the angle which r makes with the positive horizontal direction OX, the process just explained can be applied just as easily, with a small modification. Instead of a strip of area, we consider a

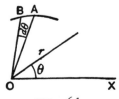

FIG. 61.

small triangle OAB, the angle at O being $d\theta$, and we find the sum of all the little triangles making up the required area.

The area of such a small triangle is approximately $\dfrac{r d\theta}{2} \times r$; hence the portion of the area included between the curve and two positions of r corresponding to the angles θ_1 and θ_2 is given by

$$\frac{1}{2} \int_{\theta=\theta_1}^{\theta=\theta_2} r^2 \, d\theta$$

Examples.

(1) Find the area of the sector of 1 radian in a circumference of radius a inch.

The polar equation of the circumference is evidently $r = a$. The area is

$$\frac{1}{2} \int_{\theta=0}^{\theta=1} a^2 \, d\theta = \frac{a^2}{2} \int_{\theta=0}^{\theta=1} d\theta = \frac{a^2}{2}$$

(2) Find the area of the first quadrant of the curve (known as a "cardioid"), the polar equation of which is

$$r = a(1 + \cos \theta)$$

$$\text{Area} = \frac{1}{2} \int_{\theta=0}^{\theta=\frac{\pi}{2}} a^2 (1 + \cos \theta)^2 d\theta$$

$$= \frac{a^2}{2} \int_{\theta=0}^{\theta=\frac{\pi}{2}} (1 + 2 \cos \theta + \cos^2 \theta) d\theta$$

$$= \frac{a^2}{2} \left[\theta + 2 \sin \theta + \frac{\theta}{2} + \frac{\sin 2\theta}{4} \right]_0^{\frac{\pi}{2}}$$

$$= \frac{a^2(3\pi + 8)}{8}$$

Volumes by Integration

What we have done with the area of a little strip of a surface, we can, of course, just as easily do with the volume of a little strip of a solid. We can add up all the little strips that make up the total solid, and find its volume, just as we have added up all the small little bits that made up an area to find the final area of the figure operated upon.

Examples.

(1) Find the volume of a sphere of radius r.

A thin spherical shell has for volume $4\pi x^2\, dx$ (see Fig. 59). Summing up all the concentric shells which make up the sphere, we have

$$\text{volume sphere} = \int_{x=0}^{x=r} 4\pi x^2\, dx = 4\pi \left[\frac{x^3}{3}\right]_0^r = \tfrac{4}{3}\pi r^3$$

We can also proceed as follows: a slice of the sphere, of thickness dx, has for volume $\pi y^2\, dx$ (see Fig. 62). Also x and y are related by the expression

$$y^2 = r^2 - x^2$$

Hence volume sphere $= 2\displaystyle\int_{x=0}^{x=r} \pi(r^2 - x^2)dx$

$$= 2\pi \left[\int_{x=0}^{x=r} r^2\, dx - \int_{x=0}^{x=r} x^2\, dx\right]$$

$$= 2\pi \left[r^2 x - \frac{x^3}{3}\right]_0^r = \frac{4\pi}{3} r^3$$

(2) Find the volume of the solid generated by the revolution of the curve $y^2 = 6x$ about the axis of x, between $x = 0$ and $x = 4$.

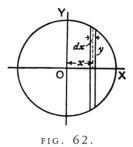

FIG. 62.

The volume of a slice of the solid is $\pi y^2 dx$.

Hence   volume $= \displaystyle\int_{x=0}^{x=4} \pi y^2 dx = 6\pi \int_{x=0}^{x=4} x\, dx$

$$= 6\pi \left[\frac{x^2}{2} \right]_0^4 = 48\pi = 150.8.$$

On Quadratic Means

In certain branches of physics, particularly in the study of alternating electric currents, it is necessary to be able to calculate the *quadratic mean* of a variable quantity. By "quadratic mean" is denoted the square root of the mean of the squares of all the values between the limits considered. Other names for the quadratic mean of any quantity are its "virtual" value, or its "R.M.S." (meaning root-mean-square) value. The French term is *valeur efficace*. If y is the function under consideration, and the quadratic mean is to be taken between the limits of $x = 0$ and $x = k$; then the quadratic mean is expressed as

$$\sqrt[2]{\frac{1}{k} \int_0^k y^2\, dx}$$

Examples.
(1) To find the quadratic mean of the function $y = ax$ (Fig. 63).

Here the intergral is $\displaystyle\int_0^k a^2 x^2 dx$ which is $\frac{1}{3}a^2 k^3$. Dividing by k

and taking the square root, we have

$$\text{quadratic mean} = \frac{1}{\sqrt{3}} ak$$

Here the arithmetical mean is $\frac{1}{2}ak$; and the ratio of quadratic to arithmetical mean (this ratio is called the *form-factor*) is $2/\sqrt{3} = 2\sqrt{3}/3 = 1.1547. \ldots$

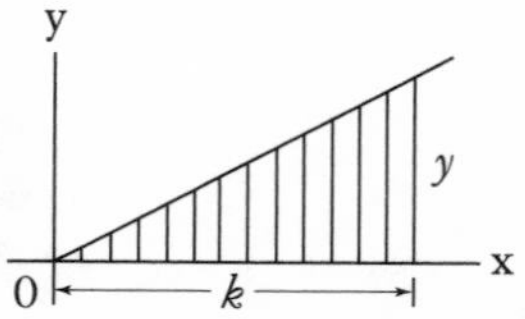

FIG. 63.

(2) To find the quadratic mean of the function $y = x^a$.

The integral is $\displaystyle\int_{x=0}^{x=k} x^{2a}dx$, that is $\dfrac{k^{2a+1}}{2a + 1}$

Hence $\qquad$ quadratic mean $= \sqrt{\dfrac{k^{2a}}{2a + 1}}$

(3) To find the quadratic mean of the function $y = a^{\frac{x}{2}}$

The integral is $\displaystyle\int_{x=0}^{x=k} \left(a^{\frac{x}{2}}\right)^2 dx$, that is $\displaystyle\int_{x=0}^{x=k} a^x dx$

or $\qquad \left[\dfrac{a^x}{\ln a}\right]_{x=0}^{x=k}$, which is $\dfrac{a^k - 1}{\ln a}$

Hence the quadratic mean is $\sqrt{\dfrac{a^k - 1}{k \ln a}}$

EXERCISES XVIII

(1) Find the area of the curve $y = x^2 + x + 5$ between $x = 0$ and $x = 6$, and the mean ordinate between these limits.

(2) Find the area of the parabola $y = 2a\sqrt{x}$ between $x = 0$ and $x = a$. Show that it is two-thirds of the rectangle of the limiting ordinate and of its abscissa.

(3) Find the area of the portion of a sine curve between $x = 0$ and $x = \pi$, and the mean ordinate.

(4) Find the area of the portion of the curve $y = \sin^2 x$ from $0°$ to $180°$, and find the mean ordinate.

(5) Find the area included between the two branches of the curve $y = x^2 \pm x^{\frac{5}{2}}$ from $x = 0$ to $x = 1$, also the area of the positive portion of the lower branch of the curve (Fig. 30).

(6) Find the volume of a cone of radius of base r, and of height h.

(7) Find the area of the curve $y = x^3 - \ln x$ between $x = 0$ and $x = 1$.

(8) Find the volume generated by the curve $y = \sqrt{1 + x^2}$, as it revolves about the axis of x, between $x = 0$ and $x = 4$.

(9) Find the volume generated by a sine curve between $x = 0$ and $x = \pi$, revolving about the axis of x.

(10) Find the area of the portion of the curve $xy = a$ included between $x = 1$ and $x = a$, where $a > 1$. Find the mean ordinate between these limits.

(11) Show that the quadratic mean of the function $y = \sin x$, between the limits of 0 and π radians, is $\dfrac{\sqrt{2}}{2}$. Find also the arithmetical mean of the same function between the same limits; and show that the form-factor is $=1.11$.

(12) Find the arithmetical and quadratic means of the function $x^2 + 3x + 2$, from $x = 0$ to $x = 3$.

(13) Find the quadratic mean and the arithmetical mean of the function $y = A_1 \sin x + A_3 \sin 3x$ between $x = 0$ and $x = 2\pi$.

(14) A certain curve has the equation $y = 3.42e^{0.21x}$. Find the area included between the curve and the axis of x, from the ordinate at $x = 2$ to the ordinate at $x = 8$. Find also the height of the mean ordinate of the curve between these points.

(15) The curve whose polar equation is $r = a(1 - \cos \theta)$ is known as the cardioid. Show that the area enclosed by the axis and the curve between $\theta = 0$ and $\theta = 2\pi$ radians is equal to 1.5 times that of the circle whose radius is a.

(16) Find the volume generated by the curve

$$y = \pm \frac{x}{6} \sqrt{x(10 - x)}$$

rotating about the axis of x.

Chapter XX

DODGES, PITFALLS, AND TRIUMPHS

Dodges.   A great part of the labor of integrating things consists in licking them into some shape that can be integrated. The books—and by this is meant the serious books—on the integral calculus are full of plans and methods and dodges and artifices for this kind of work. The following are a few of them.

Integration by Parts.   This name is given to a dodge, the formula for which is

$$\int u \, dx = ux - \int x \, du + C$$

It is useful in some cases that you can't tackle directly, for it shows that if in any case $\int x \, du$ can be found, then $\int u \, dx$ can also be found. The formula can be deduced as follows.

$$d(ux) = u \, dx + x \, du$$

which may be written

$$u \, dx = d(ux) - x \, du$$

which by direct integration gives the above expression.

Examples.

(1) Find $\int w \cdot \sin w \, dw$

Write $u = w$, and dx for $\sin w \cdot dw$. We shall then have $du = dw$, while $x = \displaystyle\int \sin w \cdot dw = -\cos w$.

Putting these into the formula, we get

$$\int w \cdot \sin w \, dw = w(-\cos w) - \int - \cos w \, dw$$

$$= -w \cos w + \sin w + C.$$

(2) Find $\displaystyle\int xe^x dx$.

Write $\qquad\qquad u = x \quad dv = e^x dx$

then $\qquad\qquad du = dx \quad v = e^x$

and $\qquad\qquad \displaystyle\int xe^x dx = xe^x - \int e^x dx$ (by the formula)

$$= xe^x - e^x + C = e^x(x - 1) + C$$

(3) Try $\displaystyle\int \cos^2 \theta \, d\theta$.

$$u = \cos\,\theta \qquad dx = \cos\,\theta\,d\theta$$

Hence $\qquad\qquad du = -\sin\,\theta\,d\theta \quad x = \sin\,\theta$

$$\int \cos^2\theta \, d\theta = \cos\,\theta\,\sin\,\theta + \int \sin^2\theta \, d\theta$$

$$= \frac{2 \cos\,\theta\,\sin\,\theta}{2} + \int (1 - \cos^2\theta)d\theta$$

$$= \frac{\sin 2\theta}{2} + \int d\theta - \int \cos^2\theta \, d\theta$$

Hence $\qquad\qquad 2\displaystyle\int \cos^2\theta \, d\theta = \frac{\sin 2\theta}{2} + \theta + 2C$

and $$\int \cos^2\theta \; d\theta = \frac{\sin 2\theta}{4} + \frac{\theta}{2} + C$$

(4) Find $\int x^2 \sin x \; dx$.

Write $$u = x^2 \qquad dv = \sin x \; dx$$

then $$du = 2x \; dx \qquad v = -\cos x$$

$$\int x^2 \sin x \; dx = -x^2 \cos x + 2 \int x \cos x \; dx$$

Now find $\int x \cos x \; dx$, integrating by parts (as in Example 1 above):

$$\int x \cos x \; dx = x \sin x + \cos x + C$$

Hence $$\int x^2 \sin x \; dx = -x^2 \cos x + 2x \sin x \; + 2 \cos x + C'$$

$$= (2 - x^2) \cos x + 2x \sin x + C'$$

(5) Find $\int \sqrt{1 - x^2} dx$.

Write $$u = \sqrt{1 - x^2}, \quad dx = dv$$

then $$du = -\frac{x \; dx}{\sqrt{1 - x^2}} \text{ (see Chap. IX)}$$

and $x = v$; so that

$$\int \sqrt{1 - x^2} dx = x \sqrt{1 - x^2} + \int \frac{x^2 dx}{\sqrt{1 - x^2}}$$

Here we may use a little dodge, for we can write

$$\int \sqrt{1-x^2}\,dx = \int \frac{(1-x^2)dx}{\sqrt{1-x^2}} = \int \frac{dx}{\sqrt{1-x^2}} - \int \frac{x^2 dx}{\sqrt{1-x^2}}$$

Adding these two last equations, we get rid of $\int \dfrac{x^2 dx}{\sqrt{1-x^2}}$,
and we have

$$2\int \sqrt{1-x^2}\,dx = x\sqrt{1-x^2} + \int \frac{dx}{\sqrt{1-x^2}}$$

Do you remember meeting $\dfrac{dx}{\sqrt{1-x^2}}$? It is got by differentiating $y = $ arc sin x; hence its integral is arc sin x, and so

$$\int \sqrt{1-x^2}\,dx = \frac{x\sqrt{1-x^2}}{2} + \tfrac{1}{2}\,\text{arc sin } x + C$$

You can try now some exercises by yourself; you will find some at the end of this chapter.

Substitution.   This is the same dodge as explained in Chap. IX. Let us illustrate its application to integration by a few examples.

(1) $\int \sqrt{3+x}\,dx$

Let $u = 3 + x, \quad du = dx$

replace: $\int u^{\frac{1}{2}}\,du = \tfrac{2}{3}u^{\frac{3}{2}} + C = \tfrac{2}{3}(3+x)^{\frac{3}{2}} + C$

(2) $\int \dfrac{dx}{e^x + e^{-x}}$

Let $\qquad u = e^x, \dfrac{du}{dx} = e^x, \text{ and } dx = \dfrac{du}{e^x}$

so that $\quad \displaystyle\int \frac{dx}{e^x + e^{-x}} = \int \frac{du}{e^x(e^x + e^{-x})} = \int \frac{du}{u\left(u + \dfrac{1}{u}\right)} = \int \frac{du}{u^2 + 1}$

$\dfrac{du}{1 + u^2}$ is the result of differentiating arc tan u.

Hence the integral is arc tan $e^x + C$.

(3) $\displaystyle\int \frac{dx}{x^2 + 2x + 3} = \int \frac{dx}{x^2 + 2x + 1 + 2} = \int \frac{dx}{(x + 1)^2 + \left(\sqrt{2}\right)^2}.$

Let $\qquad u = x + 1, \quad du = dx;$

then the integral becomes $\displaystyle\int \frac{du}{u^2 + \left(\sqrt{2}\right)^2}$; but $\dfrac{du}{u^2 + a^2}$ is the re-

sult of differentiating $\dfrac{1}{a}$ arc tan $\dfrac{u}{a}$.

Hence one has finally $\dfrac{1}{\sqrt{2}}$ arc tan $\dfrac{x + 1}{\sqrt{2}} + C$ for the value of the given integral.

Rationalization, and *Factorization of Denominator* are dodges applicable in special cases, but they do not admit of any short or general explanation. Much practice is needed to become familiar with these preparatory processes.

The following example shows how the process of splitting into partial fractions, which we learned in Chap. XIII, can be made use of in integration.

Take $\displaystyle\int \frac{dx}{x^2 + 2x - 3}$; if we split $\dfrac{1}{x^2 + 2x - 3}$ into partial fractions, this becomes:

$$\frac{1}{4}\left[\int \frac{dx}{x - 1} - \int \frac{dx}{x + 3}\right] = \frac{1}{4}\left[\ln (x - 1) - \ln (x + 3)\right] + C$$

$$= \frac{1}{4} \ln \frac{x - 1}{x + 3} + C$$

Notice that the same integral can be expressed sometimes in more than one way (which are equivalent to one another).

Pitfalls.   A beginner is liable to overlook certain points that a practised hand would avoid; such as the use of factors that are equivalent to either zero or infinity, and the occurence of indeterminate quantities such as $\frac{0}{0}$. There is no golden rule that will meet every possible case. Nothing but practice and intelligent care will avail. An example of a pitfall which had to be circumvented arose in Chap. XVIII, when we came to the problem of integrating $x^{-1}\ dx$.

Triumphs.   By triumphs must be understood the successes with which the calculus has been applied to the solution of problems otherwise intractable. Often in the consideration of physical relations one is able to build up an expression for the law governing the interaction of the parts or of the forces that govern them, such expression being naturally in the form of a *differential equation*, that is an equation containing derivatives with or without other algebraic quantities. And when such a differential equation has been found, one can get no further until it has been integrated. Generally it is much easier to state the appropriate differential equation than to solve it: the real trouble begins then only when one wants to integrate, unless indeed the equation is seen to possess some standard form of which the integral is known, and then the triumph is easy. The equation which results from integrating a differential equation is called* its "solution"; and it is quite astonishing how in many cases the solution looks as if it had no relation to the differential equation of which it is the integrated form. The solution often seems as different from the original expression as a butterfly does from a

---

*This means that the actual result of solving it is called its "solution". But many mathematicians would say, with Professor A.R. Forsyth, "every differential equation *is considered as solved* when the value of the dependent variable is expressed as a function of the independent variable by means either of known functions, or of integrals, whether the integrations in the latter can or cannot be expressed in terms of functions already known."

caterpillar that it was. Who would have supposed that such an innocent thing as

$$\frac{dy}{dx} = \frac{1}{a^2 - x^2}$$

could blossom out into

$$y = \frac{1}{2a} \ln \frac{a+x}{a-x} + C?$$

yet the latter is the *solution* of the former.

As a last example, let us work out the above together.

By partial fractions,

$$\frac{1}{a^2 - x^2} = \frac{1}{2a(a+x)} + \frac{1}{2a\,(a-x)}$$

$$dy = \frac{dx}{2a\,(a+x)} + \frac{dx}{2a\,(a-x)}$$

$$y = \frac{1}{2a} \left(\int \frac{dx}{a+x} + \int \frac{dx}{a-x} \right)$$

$$= \frac{1}{2a} \left[\ln\,(a+x) - \ln\,(a-x)\right] + C$$

$$= \frac{1}{2a} \ln \frac{a+x}{a-x} + C$$

Not a very difficult metamorphosis!

There are whole treatises, such as George Boole's *Differential Equations*, devoted to the subject of finding the "solutions" for different original forms.

EXERCISES XIX

(1) Find $\int \sqrt{a^2 - x^2}\,dx$.   (2) Find $\int x \ln x\, dx$.

(3) Find $\int x^a \ln x\, dx$.   (4) Find $\int e^x \cos e^x\, dx$.

(5) Find $\int \dfrac{1}{x} \cos (\ln x)\, dx.$ (6) Find $\int x^2 e^x\, dx.$

(7) Find $\int \dfrac{(\ln x)^a}{x}\, dx.$ (8) Find $\int \dfrac{dx}{x \ln x}.$

(9) Find $\int \dfrac{5x + 1}{x^2 + x - 2}\, dx.$ (10) Find $\int \dfrac{(x^2 - 3)dx}{x^3 - 7x + 6}.$

(11) Find $\int \dfrac{b\, dx}{x^2 - a^2}.$ (12) Find $\int \dfrac{4x\, dx}{x^4 - 1}.$

(13) Find $\int \dfrac{dx}{1 - x^4}.$ (14) Find $\int \dfrac{x\, dx}{\sqrt{a^2 - b^2 x^2}}.$

(15) Use the substitution $\dfrac{1}{x} = \dfrac{b}{a} \cosh u$ to show that

$$\int \frac{dx}{x\sqrt{a^2 - b^2 x^2}} = \frac{1}{a} \ln \frac{a - \sqrt{a^2 - b^2 x^2}}{x} + C$$

Chapter XXI

FINDING SOLUTIONS

In this chapter we go to work finding solutions to some important differential equations, using for this purpose the processes shown in the preceding chapters.

The beginner, who now knows how easy most of those processes are in themselves, will here begin to realize that integration is *an art*. As in all arts, so in this, facility can be acquired only by diligent and regular practice. Those who would attain that facility must work out examples, and more examples, and yet more examples, such as are found abundantly in all the regular treatises on the calculus. Our purpose here must be to afford the briefest introduction to serious work.

Example 1.   Find the solution of the differential equation

$$ay + b\frac{dy}{dx} = 0$$

Transposing, we have

$$b\frac{dy}{dx} = -ay$$

Now the mere inspection of this relation tells us that we have got to do with a case in which $\frac{dy}{dx}$ is proportional to y. If we think of the curve which will represent y as a function of x, it will be such that its slope at any point will be proportional to the ordinate at that point, and will be a negative slope if y is positive. So obviously the curve will be a die-away curve, and the solution will contain e^{-x} as a factor. But, without presuming on this bit of sagacity, let us go to work.

As both y and dy occur in the equation and on opposite sides, we can do nothing until we get both y and dy to one side, and dx to the other. To do this, we must split our usually inseparable companions dy and dx from one another.

$$\frac{dy}{y} = -\frac{a}{b}\,dx$$

Having done the deed, we now can see that both sides have got into a shape that is integrable, because we recognize $\frac{dy}{y}$, or $\frac{1}{y}\,dy$, as a differential that we have met with when differentiating logarithms. So we may at once write down the instructions to integrate,

$$\int \frac{dy}{y} = \int -\frac{a}{b}dx$$

and doing the two integrations, we have:

$$\ln y = -\frac{a}{b}x + \ln C$$

where $\ln C$ is the yet undetermined constant* of integration. Then, delogarizing, we get:

$$y = Ce^{-\frac{a}{b}x}$$

which is *the solution* required. Now, this solution looks quite unlike the original differential equation from which it was constructed: yet to an expert mathematician they both convey the same information as to the way in which y depends on x.

---

*We may write down any form of constant as the "constant of integration", and the form $\ln C$ is adopted here by preference, because the other terms in this line of equation are, or are treated as logarithms; and it saves complications afterward if the added constant be *of the same kind.*

Now, as to the C, its meaning depends on the initial value of y. For if we put $x = 0$ in order to see what value y then has, we find that this makes $y = Ce^{-0}$; and as $e^{-0} = 1$, we see that C is nothing else than the particular value* of y at starting. This we may call y_0 and so write the solution as

$$y = y_0 e^{-\frac{a}{b} x}$$

Example 2.
Let us take as an example to solve

$$ay + b\frac{dy}{dx} = g$$

where g is a constant. Again, inspecting the equation will suggest, (1) that somehow or other e^x will come into the solution, and (2) that if at any part of the curve y becomes either a maximum or a minimum, so that $\frac{dy}{dx} = 0$, then y will have the value = $\frac{g}{a}$ But let us go to work as before, separating the differentials and trying to transform the thing into some integrable shape.

$$b\frac{dy}{dx} = g - ay$$

$$\frac{dy}{dx} = \frac{a}{b}\left(\frac{g}{a} - y\right)$$

$$\frac{dy}{y - \frac{g}{a}} = -\frac{a}{b}dx$$

Now we have done our best to get nothing but y and dy on one side, and nothing but dx on the other. But is the result on the left side integrable?

*Compare what was said about the "constant of integration", with reference to Fig. 48, and Fig. 51.

It is of the same form as the result in Chapter XIV, so, writing the instructions to integrate, we have:

$$\int \frac{dy}{y - \frac{g}{a}} = - \int \frac{a}{b} dx$$

and, doing the integration, and adding the appropriate constant,

$$\ln \left(y - \frac{g}{a} \right) = -\frac{a}{b} x + \ln C$$

whence
$$y - \frac{g}{a} = Ce^{-\frac{a}{b} x}$$

and finally,
$$y = \frac{g}{a} + Ce^{-\frac{a}{b} x}$$

which is *the solution*.

If the condition is laid down that $y = 0$ when $x = 0$ we can find C; for then the exponential becomes $=1$; and we have

$$0 = \frac{g}{a} + C$$

or
$$C = -\frac{g}{a}$$

Putting in this value, the solution becomes

$$y = \frac{g}{a}(1 - e^{-\frac{a}{b} x})$$

But further, if x grows infinitely, y will grow to a maximum; for when $x = \infty$, the exponential $= 0$, giving $y_{\text{max.}} = \frac{g}{a}$. Substituting this, we get finally

$$y = y_{\text{max.}}(1 - e^{-\frac{a}{b} x})$$

This result is also of importance in physical science.

Before proceeding to the next example, it is necessary to discuss two integrals which are of great importance in physics and engineering. These seem to be very elusive as, when either of them is tackled, it turns partly into the other. Yet this very fact helps us to determine their values. Let us denote these integrals by S and C, where

$$S = \int e^{pt} \sin kt \, dt, \quad \text{and} \quad C = \int e^{pt} \cos kt \, dt,$$

where p and k are constants.

To tackle these formidable-looking integrals, we resort to the device of integrating by parts, the general formula of which is

$$\int u \, dv = uv - \int v \, du$$

For this purpose, write $u = e^{pt}$ and $dv = \sin kt \, dt$ in S; then $du = pe^{pt} \, dt$, and $v = \int \sin kt \, dt = -\dfrac{1}{k} \cos kt$, omitting temporarily the constant.

Inserting these values, the integral S becomes

$$S = \int e^{pt} \sin kt \, dt = -\frac{1}{k} e^{pt} \cos kt - \int -\frac{1}{k} \cos kt \, pe^{pt} \, dt$$

$$= -\frac{1}{k} e^{pt} \cos kt + \frac{p}{k} \int e^{pt} \cos kt \, dt$$

$$= -\frac{1}{k} e^{pt} \cos kt + \frac{p}{k} C \, \ldots\ldots\ldots\ldots\ldots (i)$$

Thus the dodge of integrating by parts turns S partly into C. But let us look at C. Writing $u = e^{pt}$, as before, $dv = \cos kt \, dt$, then $v = \dfrac{1}{k} \sin kt$; hence, the rule for integrating by parts gives

$$C = \int e^{pt} \cos kt \, dt = \frac{1}{k} e^{pt} \sin kt - \frac{p}{k} \int e^{pt} \sin kt \, dt$$

$$= \frac{1}{k} e^{pt} \sin kt - \frac{p}{k} S. \, \ldots\ldots\ldots\ldots (ii)$$

The facts that S turns partly into C, and C partly into S might lead you to think that the integrals are intractable, but from the relations (i) and (ii), which may be regarded as two equations in S and C, the integrals themselves may be readily deduced.

Thus, substitute in (i) the value of C from (ii), then

$$S = -\frac{1}{k}\, e^{pt} \cos kt + \frac{p}{k} \left(\frac{1}{k}\, e^{pt} \sin kt - \frac{p}{k} S \right)$$

or

$$S \left(\frac{p^2}{k^2} + 1 \right) = \frac{1}{k^2}\, e^{pt} \left(p \sin kt - k \cos kt \right)$$

from which

$$S = \frac{e^{pt}}{p^2 + k^2} \left(p \sin kt - k \cos kt \right)$$

The integral C may be obtained in like manner by inserting in (ii) the equivalent of S given by (i); the final result is

$$C = \frac{e^{pt}}{p^2 + k^2} \left(p \cos kt + k \sin kt \right)$$

We have, therefore, the following very important integrals to add to our list, namely:

$$\int e^{pt} \sin kt\, dt = \frac{e^{pt}}{p^2 + k^2} \left(p \sin kt - k \cos kt \right) + E$$

$$\int e^{pt} \cos kt\, dt = \frac{e^{pt}}{p^2 + k^2} \left(p \cos kt + k \sin kt \right) + F$$

where E and F are the constants of integration.

Example 3.

Let

$$ay + b\frac{dy}{dt} = g \sin 2\pi nt$$

First divide through by b.

$$\frac{dy}{dt} + \frac{a}{b} y = \frac{g}{b} \sin 2\pi nt$$

Now, as it stands, the left side is not integrable. But it can be made so by the artifice—and this is where skill and practice suggest a plan—of multiplying all the terms by $e^{\frac{a}{b}t}$, giving us:

$$\frac{dy}{dt}e^{\frac{a}{b}t} + \frac{a}{b}ye^{\frac{a}{b}t} = \frac{g}{b}e^{\frac{a}{b}t}\sin 2\pi nt$$

For if $u = ye^{\frac{a}{b}t}$, $\dfrac{du}{dt} = \dfrac{dy}{dt}e^{\frac{a}{b}t} + \dfrac{a}{b}ye^{\frac{a}{b}t}$

The equation thus becomes

$$\frac{du}{dt} = \frac{g}{b}e^{\frac{a}{b}t}\sin 2\pi nt$$

Hence, integrating gives

$$u \text{ or } ye^{\frac{a}{b}t} = \frac{g}{b}\int e^{\frac{a}{b}t}\sin 2\pi nt\, dt + K$$

But the right-hand integral is of the same form as S which has just been evaluated; hence putting $p = \dfrac{a}{b}$ and $k = 2\pi n$;

$$ye^{\frac{a}{b}t} = \frac{ge^{\frac{a}{b}t}}{a^2 + 4\pi^2n^2b^2}(a\sin 2\pi nt - 2\pi nb\cos 2\pi nt) + K$$

or

$$y = g\left\{\frac{a\sin 2\pi nt - 2\pi nb\cos 2\pi nt}{a^2 + 4\pi^2n^2b^2}\right\} + Ke^{-\frac{a}{b}t}$$

To simplify still further, let us imagine an angle ϕ such that $\tan \phi = 2\pi nb/a$.

Then $\sin \phi = \dfrac{2\pi nb}{\sqrt{a^2 + 4\pi^2n^2b^2}}$, and $\cos \phi = $

$\dfrac{a}{\sqrt{a^2 + 4\pi^2n^2b^2}}$. Substituting these, we get:

$$y = g\frac{\cos \phi \sin 2\pi nt - \sin \phi \cos 2\pi nt}{\sqrt{a^2 + 4\pi^2n^2b^2}}$$

or
$$y = g\frac{\sin(2\pi n t - \phi)}{\sqrt{a^2 + 4\pi^2 n^2 b^2}}$$

which is *the solution* desired, omitting the constant which dies out.

This is indeed none other than the equation of an alternating electric current, where g represents the amplitude of the electromotive force, n the frequency, a the resistance, b the coefficient of induction of the circuit, and ϕ is the delay of a phase angle.

Example 4.

Suppose that $\qquad M\,dx + N\,dy = 0$

We could integrate this expression directly, if M were a function of x only, and N a function of y only; but, if both M and N are functions that depend on both x and y, how are we to integrate it? Is it itself an exact differential? That is: have M and N each been formed by partial differentiations from some common function U, or not? If they have, then

$$\frac{\partial U}{\partial x} = M, \text{ and } \frac{\partial U}{\partial y} = N$$

And if such a common function exists, then

$$\frac{\partial U}{\partial x}dx + \frac{\partial U}{\partial y}dy$$

is an exact differential.

Now the test of the matter is this. If the expression is an exact differential, it must be true that

$$\frac{\delta M}{\delta y} = \frac{\delta N}{\delta x}$$

for then
$$\frac{\delta(\delta U)}{\delta x\,\delta y} = \frac{\delta(\delta U)}{\delta y\,\delta x}$$

which is necessarily true.

Take as an illustration the equation

$$(1 + 3xy)dx + x^2 dy = 0$$

Is this an exact differential or not? Apply the test.

$$\frac{\delta(1 + 3xy)}{\delta y} = 3x \qquad \frac{\delta(x^2)}{\delta x} = 2x$$

which do not agree. Therefore, it is not an exact differential, and the two functions $1 + 3xy$ and x^2 have not come from a common original function.

It is possible in such cases to discover, however, *an integrating factor,* that is to say, a factor such that if both are multiplied by this factor, the expression will become an exact differential. There is no one rule for discovering such an integrating factor; but experience will usually suggest one. In the present instance $2x$ will act as such. Multiplying by $2x$, we get

$$(2x + 6x^2y)dx + 2x^3dy = 0$$

Now apply the test to this.

$$\frac{\delta(2x + 6x^2y)}{\delta y} = 6x^2 \qquad \frac{\delta(2x^3)}{\delta x} = 6x^2$$

which agrees. Hence this is an exact derivative, and may be integrated. Now, if $w = 2x^3y$,

$$dw = 6x^2y \, dx + 2x^3dy$$

Hence

$$\int 6x^2y \, dx + \int 2x^3 \, dy = w = 2x^3y$$

so that we get

$$U = x^2 + 2x^3y + C$$

Example 5. Let $\dfrac{d^2y}{dt^2} + n^2\, y = 0$

In this case we have a differential equation of the second degree, in which y appears in the form of a second derivative, as well as in person. Transposing, we have

$$\frac{d^2y}{dt^2} = -n^2y$$

It appears from this that we have to do with a function such that its second derivative is proportional to itself, but with reversed sign. In Chapter XV we found that there was such a function—namely, the *sine* (or the *cosine* also) which possessed this property. So, without further ado, we may guess that the solution will be of the form

$$y = A \sin (pt + q)$$

However, let us go to work.

Multiply both sides of the original equation by $2\dfrac{dy}{dt}$ and integrate, giving us $2\dfrac{d^2y}{dt^2}\dfrac{dy}{dt} + 2n^2y\dfrac{dy}{dt} = 0$, and, as

$$2\frac{d^2y}{dt^2}\frac{dy}{dt} = \frac{d\left(\dfrac{dy}{dt}\right)^2}{dt}, \quad \left(\frac{dy}{dt}\right)^2 + n^2(y^2 - C^2) = 0$$

C being a constant. Then, taking the square roots,

$$\frac{dy}{dt} = n\sqrt{C^2 - y^2} \quad \text{and} \quad \frac{dy}{\sqrt{C^2 - y^2}} = n \cdot dt$$

But it can be shown that

$$\frac{1}{\sqrt{C^2 - y^2}} = \frac{d\left(\arcsin\dfrac{y}{C}\right)}{dy}$$

whence, passing from angles to sines,

$$\arcsin \frac{y}{C} = nt + C_1 \text{ and } y = C \sin (nt + C_1)$$

where C_1 is a constant angle that comes in by integration.
Or, preferably, this may be written

$$y = A \sin nt + B \cos nt, \text{ which is the solution.}$$

Example 6. $$\frac{d^2y}{dx^2} - n^2y = 0$$

Here we have obviously to deal with a function y which is such that its second derivative is proportional to itself. The only function we know that has this property is the exponential function, and we may be certain therefore that the solution of the equation will be of that form.

Proceeding as before, by multiplying through by $2\frac{dy}{dx}$, and integrating, we get $2\frac{d^2y}{dx^2}\frac{dy}{dx} - 2n^2y\frac{dy}{dx} = 0$, and, as

$$2\frac{d^2y}{dx^2}\frac{dy}{dx} = \frac{d\left(\frac{dy}{dx}\right)^2}{dx}, \left(\frac{dy}{dx}\right)^2 - n^2(y^2 + c^2) = 0$$

$$\frac{dy}{dx} - n\sqrt{y^2 + c^2} = 0$$

where c is a constant, and $\dfrac{dy}{\sqrt{y^2 + c^2}} = n\,dx$.

To integrate this equation it is simpler to use hyperbolic functions.

Let $y = c \sinh u$, then $dy = c \cosh u\,du$, and

$$y^2 + c^2 = c^2(\sinh^2 u + 1) = c^2 \cosh^2 u.$$

$$\int \frac{dy}{\sqrt{y^2 + c^2}} = \int \frac{c \cosh u\,du}{c \cosh u} = \int du = u$$

Hence, the integral of the equation

$$n \int dx = \int \frac{dy}{\sqrt{y^2 + c^2}}$$

is $$nx + K = u$$

where K is the constant of integration, and $c \sinh u = y$.

$$\sinh (nx + K) = \sinh u = \frac{y}{c}$$

or
$$y = c \sinh (nx + K)$$
$$= \tfrac{1}{2}c(e^{nx+K} - e^{-nx-K})$$
$$= Ae^{nx} + Be^{-nx}$$

where $A = \tfrac{1}{2}ce^{K}$ and $B = -\tfrac{1}{2}ce^{-K}$.

This solution which at first sight does not look as if it had anything to do with the original equation, shows that y consists of two terms, one of which grows exponentially as x increases, while the other term dies away as x increases.

Example 7.

Let
$$b\frac{d^2y}{dt^2} + a\frac{dy}{dt} + gy = 0.$$

Examination of this expression will show that, if $b = 0$, it has the form of Example 1, the solution of which was a negative exponential. On the other hand, if $a = 0$, its form becomes the same as that of Example 6, the solution of which is the sum of a positive and a negative exponential. It is therefore not very surprising to find that the solution of the present example is

$$y = (e^{-mt})(Ae^{nt} + Be^{-nt})$$

where
$$m = \frac{a}{2b} \quad \text{and} \quad n = \frac{\sqrt{a^2 - 4bg}}{2b}$$

The steps by which this solution is reached are not given here; they may be found in advanced treatises.

Example 8.

$$\frac{\delta^2 y}{\delta t^2} = a^2 \frac{\delta^2 y}{\delta x^2}$$

It was seen earlier that this equation was derived from the original

$$y = F(x + at) + f(x - at)$$

where F and f were any arbitrary functions of t.

Another way of dealing with it is to transform it by a change of variables into

$$\frac{\delta^2 y}{\delta u \cdot \delta v} = 0$$

where $u = x + at$, and $v = x - at$, leading to the same general solution. If we consider a case in which F vanishes, then we have simply

$$y = f(x - at)$$

and this merely states that, at the time $t = 0$, y is a particular function of x, and may be looked upon as denoting that the curve of the relation of y to x has a particular shape. Then any change in the value of t is equivalent simply to an alteration in the origin from which x is reckoned. That is to say, it indicates that, the form of the function being conserved, it is propagated along the x direction with a uniform velocity a; so that whatever the value of the ordinate y at any particular time t_0 at any particular point x_0, the same value of y will appear at the subsequent time t_1 at a point further along, the abscissa of which is $x_0 + a(t_1 - t_0)$. In this case the simplified equation represents the propagation of a wave (of any form) at a uniform speed along the x direction.

If the differential equation had been written

$$m \frac{d^2 y}{dt^2} = k \frac{d^2 y}{dx^2}$$

the solution would have been the same, but the velocity of propagation would have had the value

$$a = \sqrt{\frac{k}{m}}$$

EXERCISES XX

Try to solve the following equations.

(1) $\dfrac{dT}{d\theta} = \mu T$, given that μ is constant, and when $\theta = 0$, $T = T_0$.

(2) $\dfrac{d^2s}{dt^2} = a$, where a is constant. When $t = 0$, $s = 0$ and $\dfrac{ds}{dt} = u$.

(3) $\dfrac{di}{dt} + 2i = \sin 3t$, it being known that $i = 0$ when $t = 0$.

(*Hint.* Multiply out by e^{2t}.)

Chapter XXII

A LITTLE MORE ABOUT CURVATURE OF CURVES

In Chapter XII we have learned how we can find out which way a curve is curved, that is, whether it curves upwards or downwards towards the right. This gave us no indication whatever as to *how much* the curve is curved, or, in other words, what is its *curvature*.

By *curvature* of a curve, we mean the amount of bending or deflection taking place along a certain length of the curve, say along a portion of the curve the length of which is one unit of length (the same unit which is used to measure the radius, whether it be one inch, one foot, or any other unit). For instance, consider two circular paths of center O or O' and of equal lengths AB, $A'B'$ (see Fig. 64). When passing from A to B along the arc AB of the first one, one changes one's direction from AP to BQ, since at A

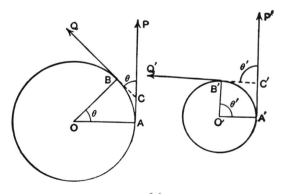

FIG. 64.

one faces in the direction AP and at B one faces in the direction BQ. In other words, in walking from A to B one unconsciously turns round through the angle PCQ, which is equal to the angle AOB. Similarly, in passing from A' to B', along the arc $A'B'$, of equal length to AB, on the second path, one turns round through the angle $P'C'Q'$, which is equal to the angle $A'O'B'$, obviously, *greater* than the corresponding angle AOB. The second path bends therefore more than the first for an equal length.

This fact is expressed by saying that the *curvature* of the second path is greater than that of the first one. The larger the circle, the lesser the bending, that is, the lesser the curvature. If the radius of the first circle is $2, 3, 4, \ldots$ etc. times greater than the radius of the second, then the angle of bending or deflection along an arc of unit length will be $2, 3, 4, \ldots$ etc. times less on the first circle than on the second, that is, it will be $\frac{1}{2}, \frac{1}{3}, \frac{1}{4}, \ldots$ etc. of the bending or deflection along the arc of same length on the second circle. In other words, the *curvature* of the first circle will be $\frac{1}{2}, \frac{1}{3}, \frac{1}{4}, \ldots$ etc. of that of the second circle. We see that, as the radius becomes $2, 3, 4, \ldots$ etc. times greater, the curvature becomes $2, 3, 4, \ldots$ etc. times smaller, and this is expressed by saying that *the curvature of a circle is inversely proportional to the radius of the circle,* or

$$\text{curvature} = k \times \frac{1}{\text{radius}}$$

where k is a constant. It is agreed to take $k = 1$, so that

$$\text{curvature} = \frac{1}{\text{radius}}$$

always.

If the radius becomes infinitely great, the curvature becomes $\frac{1}{\text{infinity}} = $ zero, since when the denominator of a fraction is infinitely large, the value of the fraction is infinitely small. For this reason mathematicians sometimes consider a straight line as an arc of circle of infinite radius, or zero curvature.

In the case of a circle, which is perfectly symmetrical and uniform, so that the curvature is the same at every point of its circumference, the above method of expressing the curvature is perfectly definite. In the case of any other curve, however, the curvature is not the same at different points, and it may differ considerably even for two points fairly close to one another. It would not then be accurate to take the amount of bending or deflection between two points as a measure of the curvature of the arc between these points, unless this arc is very small, in fact, unless it is infinitely small.

If then we consider a very small arc such as AB (see Fig. 65), and if we draw such a circle that an arc AB of this circle coincides with the arc AB of the curve more closely than would be the case with any other circle, then the curvature of this circle may be taken as the curvature of the arc AB of the curve. The smaller the arc AB, the easier it will be to find a circle an arc of which most nearly coincides with the arc AB of the curve. When A and B are very near one another, so that AB is so small so that the length ds of the arc AB is practically negligible, then the coincidence of the two arcs, of circle and of curve, may be considered as being practically perfect, and the curvature of the curve at the point A (or B), being then the same as the curvature of the circle, will be expressed by the reciprocal of the radius of this circle, that is, by $\frac{1}{OA}$, according to our way of measuring curvature, explained above.

Now, at first, you may think that, if AB is very small, then the circle must be very small also. A little thinking will, however,

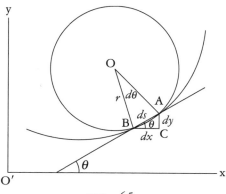

FIG. 65.

cause you to perceive that it is by no means necessarily so, and
that the circle may have any size, according to the amount of
bending of the curve along this very small arc AB. In fact, if the
curve is almost flat at that point, the circle will be extremely
large. This circle is called the *circle of curvature,* or the *osculating
circle* at the point considered. Its radius is the *radius of curvature* of
the curve at that particular point.

If the arc AB is represented by ds and the angle AOB by $d\theta$,
then, if r is the radius of curvature,

$$ds = r\,d\theta \quad \text{or} \quad \frac{d\theta}{ds} = \frac{1}{r}$$

The secant AB makes with the axis OX the angle θ, and it will
be seen from the small triangle ABC that $\dfrac{dy}{dx} = \tan\theta$. When AB
is infinitely small, so that B practically coincides with A, the line
AB becomes a tangent to the curve at the point A (or B).

Now, $\tan\theta$ depends on the position of the point A (or B, which
is supposed to nearly coincide with it), that is, it depends on x,
or, in other words, $\tan\theta$ is "a function" of x.

Differentiating with regard to x to get the slope, we get

$$\frac{d\left(\dfrac{dy}{dx}\right)}{dx} = \frac{d(\tan\theta)}{dx} \quad \text{or} \quad \frac{d^2y}{dx^2} = \sec^2\theta\,\frac{d\theta}{dx} = \frac{1}{\cos^2\theta}\frac{d\theta}{dx}$$

hence
$$\frac{d\theta}{dx} = \cos^2\theta\,\frac{d^2y}{dx^2}$$

But $\dfrac{dx}{ds} = \cos\theta$, and for $\dfrac{d\theta}{ds}$ one may write $\dfrac{d\theta}{dx} \times \dfrac{dx}{ds}$; therefore

$$\frac{1}{r} = \frac{d\theta}{ds} = \frac{d\theta}{dx} \times \frac{dx}{ds} = \cos^3\theta\,\frac{d^2y}{dx^2} = \frac{\dfrac{d^2y}{dx^2}}{\sec^3\theta}$$

but $\sec\theta = \sqrt{1 + \tan^2\theta}$; hence

$$\frac{1}{r} = \frac{\dfrac{d^2y}{dx^2}}{\left(\sqrt{1 + \tan^2\theta}\right)^3} = \frac{\dfrac{d^2y}{dx^2}}{\left\{1 + \left(\dfrac{dy}{dx}\right)^2\right\}^{\frac{3}{2}}}$$

and finally,

$$r = \frac{\left\{1 + \left(\dfrac{dy}{dx}\right)^2\right\}^{\frac{3}{2}}}{\dfrac{d^2y}{dx^2}}$$

The numerator, being a square root, may have the sign + or the sign −. One must select for it the same sign as the denominator, so as to have r positive always, as a negative radius would have no meaning.[1]

It has been shown (Chapter XII) that if $\dfrac{d^2y}{dx^2}$ is positive, the curve is concave upword, while if $\dfrac{d^2y}{dx^2}$ is negative, the curve is concave downward. If $\dfrac{d^2y}{dx^2} = 0$, the radius of curvature is infinitely great, that is, the corresponding portion of the curve is a bit of straight line. This necessarily happens whenever a curve gradually changes from being convex to concave to the axis of x or vice versa. The point where this occurs is called a *point of inflection*.

The center of the circle of curvature is called the *center of curvature*. If its coordinates are x_1, y_1, then the equation of the circle is

$$(x - x_1)^2 + (y - y_1)^2 = r^2$$

hence

$$2(x - x_1)dx + 2(y - y_1)dy = 0$$

and

$$x - x_1 + (y - y_1)\frac{dy}{dx} = 0 \quad ....(1)$$

1. Thompson is not always clear as to when a square root is to be taken as positive. In modern textbooks $\sqrt{x}$ (or in this case $x^{\frac{3}{2}}$) means the positive value. If we want to allow either the positive or the negative value we write $\pm\sqrt{x}$. The formula for the radius of curvature is usually written with $\left|\dfrac{d^2y}{dx^2}\right|$ in the denominator; since the numerator is positive, this makes r positive.—M.G.

Why did we differentiate? To get rid of the constant r. This leaves but two unknown constants x_1 and y_1; differentiate again; you shall get rid of one of them. This last differentiation is not quite as easy as it seems: let us do it together; we have:

$$\frac{d(x)}{dx} + \frac{d\left[(y-y_1)\dfrac{dy}{dx}\right]}{dx} = 0$$

the numerator of the second term is a product; hence differentiating it gives

$$(y-y_1)\frac{d\left(\dfrac{dy}{dx}\right)}{dx} + \frac{dy}{dx}\frac{d(y-y_1)}{dx} = (y-y_1)\frac{d^2y}{dx^2} + \left(\frac{dy}{dx}\right)^2$$

so that the result of differentiating (1) is

$$1 + \left(\frac{dy}{dx}\right)^2 + (y-y_1)\frac{d^2y}{dx^2} = 0$$

from this we at once get

$$y_1 = y + \frac{1 + \left(\dfrac{dy}{dx}\right)^2}{\dfrac{d^2y}{dx^2}}$$

Replacing in (1), we get

$$(x-x_1) + \left\{ y - y - \frac{1 + \left(\dfrac{dy}{dx}\right)^2}{\dfrac{d^2y}{dx^2}} \right\} \frac{dy}{dx} = 0$$

and finally,

$$x_1 = x - \frac{\dfrac{dy}{dx}\left\{1 + \left(\dfrac{dy}{dx}\right)^2\right\}}{\dfrac{d^2y}{dx^2}}$$

x_1 and y_1 give the position of the center of curvature. The use of these formulae will be best seen by carefully going through a few worked-out examples.

Example 1. Find the radius of curvature and the coordinates of the center of curvature of the curve $y = 2x^2 - x + 3$ at the point $x = 0$.

We have
$$\frac{dy}{dx} = 4x - 1 \qquad \frac{d^2y}{dx^2} = 4$$

$$r = \frac{\pm\left\{1 + \left(\dfrac{dy}{dx}\right)^2\right\}^{\frac{3}{2}}}{\dfrac{d^2y}{dx^2}} = \frac{\{1 + (4x - 1)^2\}^{\frac{3}{2}}}{4}$$

when $x = 0$; this becomes
$$\frac{\{1 + (-1)^2\}^{\frac{3}{2}}}{4} = \frac{\sqrt{8}}{4} = 0.707$$

If x_1, y_1 are the coordinates of the center of curvature, then
$$x_1 = x - \frac{\dfrac{dy}{dx}\left\{1 + \left(\dfrac{dy}{dx}\right)^2\right\}}{\dfrac{d^2y}{dx^2}} = x - \frac{(4x - 1)\{1 + (4x - 1)^2\}}{4}$$

$$= 0 - \frac{(-1)\{1 + (-1)^2\}}{4} = \frac{1}{2}$$

when $x = 0$, $y = 3$, so that
$$y_1 = y + \frac{1 + \left(\dfrac{dy}{dx}\right)^2}{\dfrac{d^2y}{dx^2}} = y + \frac{1 + (4x - 1)^2}{4} = 3 + \frac{1 + (-1)^2}{4} = 3\tfrac{1}{2}$$

Plot the curve and draw the circle; it is both interesting and instructive. The values can be checked easily, as since when $x = 0$, $y = 3$, here
$$x_1{}^2 + (y_1 - 3)^2 = r^2 \quad \text{or} \quad 0.5^2 + 0.5^2 = 0.5 = 0.707^2$$

Example 2.   Find the radius of curvature and the position of the center of curvature of the curve $y^2 = mx$ at the point for which $y = 0$.

Here $$y = m^{\frac{1}{2}}x^{\frac{1}{2}}, \quad \frac{dy}{dx} = \frac{1}{2}m^{\frac{1}{2}}x^{-\frac{1}{2}} = \frac{m^{\frac{1}{2}}}{2x^{\frac{1}{2}}}$$

$$\frac{d^2y}{dx^2} = -\frac{1}{2} \times \frac{m^{\frac{1}{2}}}{2} x^{-\frac{3}{2}} = -\frac{m^{\frac{1}{2}}}{4x^{\frac{3}{2}}}$$

hence $$\frac{\pm\left\{1 + \left(\dfrac{dy}{dx}\right)^2\right\}^{\frac{3}{2}}}{\dfrac{d^2y}{dx^2}} = \frac{\pm\left\{1 + \dfrac{m}{4x}\right\}^{\frac{3}{2}}}{-\dfrac{m^{\frac{1}{2}}}{4x^{\frac{3}{2}}}} = \frac{(4x+m)^{\frac{3}{2}}}{2m^{\frac{1}{2}}}$$

taking the − sign at the numerator, so as to have r positive.

Since, when $y = 0$, $x = 0$, we get $r = \dfrac{m^{\frac{3}{2}}}{2m^{\frac{1}{2}}} = \dfrac{m}{2}$

Also, if x_1, y_1 are the coordinates of the center,

$$x_1 = x - \frac{\dfrac{dy}{dx}\left\{1 + \left(\dfrac{dy}{dx}\right)^2\right\}}{\dfrac{d^2y}{d^2x}} = x - \frac{\dfrac{m^{\frac{1}{2}}}{2x^{\frac{1}{2}}}\left\{1 + \dfrac{m}{4x}\right\}}{-\dfrac{m^{\frac{1}{2}}}{4x^{\frac{3}{2}}}}$$

$$= x + \frac{4x + m}{2} = 3x + \frac{m}{2}$$

when $x = 0$, then $x_1 = \dfrac{m}{2}$

Also $$y_1 = y + \frac{1 + \left(\dfrac{dy}{dx}\right)^2}{\dfrac{d^2y}{dx^2}} = m^{\frac{1}{2}}x^{\frac{1}{2}} - \frac{1 + \dfrac{m}{4x}}{\dfrac{m^{\frac{1}{2}}}{4x^{\frac{3}{2}}}} = -\frac{4x^{\frac{3}{2}}}{m^{\frac{1}{2}}}$$

when $x = 0$, $y_1 = 0$.

Example 3.   Show that the circle is a curve of constant curvature.

If x_1, y_1 are the coordinates of the center, and R is the radius, the equation of the circle in rectangular coordinates is

$$(x - x_1)^2 + (y - y_1)^2 = R^2$$

Let $x - x_1 = R \cos \theta$, then

$$(y - y_1)^2 = R^2 - R^2 \cos^2\theta = R^2(1 - \cos^2\theta) = R^2 \sin^2\theta$$

$$y - y_1 = R \sin \theta$$

R, θ are thus the polar coordinates of any point on the circle referred to its center as pole.

Since $x - x_1 = R \cos \theta$, and $y - y_1 = R \sin \theta$,

$$\frac{dx}{d\theta} = -R \sin \theta, \quad \frac{dy}{d\theta} = R \cos \theta$$

$$\frac{dy}{dx} = \frac{dy}{d\theta} \cdot \frac{d\theta}{dx} = -\cot \theta$$

Further, $\dfrac{d^2y}{dx^2} = -(-\csc^2\theta) \dfrac{d\theta}{dx} = \csc^2\theta \cdot \left(-\dfrac{\csc \theta}{R}\right)$

$$= -\frac{\csc^3 \theta}{R}. \quad \text{(See Ex. 5, \textit{Chapter XV.})}$$

Hence $r = \dfrac{\pm(1 + \cot^2 \theta)^{\frac{3}{2}}}{-\dfrac{\csc^3\theta}{R}} = \dfrac{R \csc^3\theta}{\csc^3\theta} = R$

Thus the radius of curvature is constant and equal to the radius of the circle.

Example 4.   Find the radius of curvature of the curve $x = 2 \cos^3 t$, $y = 2 \sin^3 t$ at any point (x, y).

Here $dx = -6 \cos^2 t \sin t \, dt$ (see Ex. 2, Chapter XV)

and $dy = 6 \sin^2 t \cos t \, dt.$

$$\frac{dy}{dx} = -\frac{6 \sin^2 t \cos t \, dt}{6 \sin t \cos^2 t \, dt} = -\frac{\sin t}{\cos t} = -\tan t$$

Hence $\dfrac{d^2y}{dx^2} = \dfrac{d}{dt}(-\tan t)\dfrac{dt}{dx} = \dfrac{-\sec^2 t}{-6\cos^2 t \sin t} = \dfrac{\sec^4 t}{6 \sin t}$

$$r = \dfrac{\pm(1+\tan^2 t)^{\frac{3}{2}} \times 6 \sin t}{\sec^4 t} = \dfrac{6 \sec^3 t \sin t}{\sec^4 t}$$

$$= 6 \sin t \cos t = 3 \sin 2t, \text{ for } 2 \sin t \cos t = \sin 2t$$

Example 5.   Find the radius and the center of curvature of the curve $y = x^3 - 2x^2 + x - 1$ at points where $x = 0$, $x = 0.5$ and $x = 1$. Find also the position of the point of inflection of the curve.

Here $\dfrac{dy}{dx} = 3x^2 - 4x + 1, \dfrac{d^2y}{dx^2} = 6x - 4$

$$r = \dfrac{\{1 + (3x^2 - 4x + 1)^2\}^{\frac{3}{2}}}{6x - 4}$$

$$x_1 = x - \dfrac{(3x^2 - 4x + 1)\{1 + (3x^2 - 4x + 1)^2\}}{6x - 4}$$

$$y_1 = y + \dfrac{1 + (3x^2 - 4x + 1)^2}{6x - 4}$$

When $x = 0$, $y = -1$,

$$r = \dfrac{\sqrt{8}}{4} = 0.707, x_1 = 0 + \tfrac{1}{2} = 0.5, y_1 = -1 - \tfrac{1}{2} = -1.5.$$

When $x = 0.5$, $y = -0.875$:

$$r = \dfrac{-\{1 + (-0.25)^2\}^{\frac{3}{2}}}{-1} = 1.09$$

$$x_1 = 0.5 - \dfrac{-0.25 \times 1.0625}{-1} = 0.23$$

$$y_1 = -0.875 + \dfrac{1.0625}{-1} = -1.94$$

When $x = 1$, $y = -1$.

$$r = \frac{(1+0)^{\frac{3}{2}}}{2} = 0.5$$

$$x_1 = 1 - \frac{0 \times (1+0)}{2} = 1$$

$$y_1 = -1 + \frac{1 + 0^2}{2} = -0.5$$

At the point of inflection $\dfrac{d^2y}{dx^2} = 0$, $6x - 4 = 0$, and $x = \frac{2}{3}$; hence

$y = -0.926$.

Example 6.   Find the radius and center of curvature of the curve
$y = \dfrac{a}{2}\left\{ e^{\frac{x}{a}} + e^{-\frac{x}{a}} \right\}$, at the point for which $x = 0$. (This curve is called

the *catenary*, as a hanging chain affects the same slope exactly.)
The equation of the curve may be written

$$y = \frac{a}{2} e^{\frac{x}{a}} + \frac{a}{2} e^{-\frac{x}{a}}$$

then,

$$\frac{dy}{dx} = \frac{a}{2} \times \frac{1}{a} e^{\frac{x}{a}} - \frac{a}{2} \times \frac{1}{a} e^{-\frac{x}{a}} = \frac{1}{2}\left(e^{\frac{x}{a}} - e^{-\frac{x}{a}} \right)$$

Similarly

$$\frac{d^2y}{dx^2} = \frac{1}{2a}\left\{ e^{\frac{x}{a}} + e^{-\frac{x}{a}} \right\} = \frac{1}{2a} \times \frac{2y}{a} = \frac{y}{a^2}$$

$$r = \frac{\left\{ 1 + \frac{1}{4}\left(e^{\frac{x}{a}} - e^{-\frac{x}{a}} \right)^2 \right\}^{\frac{3}{2}}}{\dfrac{y}{a^2}} = \frac{a^2}{8y} \sqrt{\left(2 + e^{\frac{2x}{a}} + e^{-\frac{2x}{a}} \right)^3}$$

since $e^{\frac{x}{a}-\frac{x}{a}} = e^0 = 1$, or

$$r = \frac{a^2}{8y} \sqrt{\left(2e^{\frac{x}{a}-\frac{x}{a}} + e^{\frac{2x}{a}} + e^{-\frac{2x}{a}}\right)^3} = \frac{a^2}{8y} \sqrt{\left(e^{\frac{x}{a}} + e^{-\frac{x}{a}}\right)^6} = \frac{y^2}{a}$$

when $\qquad x = 0,\ y = \frac{a}{2}(e^0 + e^0) = a$

hence $\qquad\qquad r = \frac{a^2}{a} = a$

The radius of curvature at the vertex is equal to the constant a.

Also when $\quad x = 0,\ x_1 = 0 - \dfrac{0(1+0)}{\dfrac{1}{a}} = 0$

and $\qquad\qquad y_1 = y + \dfrac{1+0}{\dfrac{1}{a}} = a + a = 2a$

As defined earlier,

$$\frac{1}{2}\left(e^{\frac{x}{a}} + e^{-\frac{x}{a}}\right) = \cosh\frac{x}{a}$$

thus the equation of the catenary may be written in the form

$$y = a \cosh\frac{x}{a}$$

It will therefore be a useful exercise for you to verify the above results from this form of the equation.

You are now sufficiently familiar with this type of problem to work out the following exercises by yourself. You are advised to check your answers by careful plotting of the curve and construction of the circle of curvature, as explained in Example 4.

EXERCISES XXI

(1) Find the radius of curvature and the position of the center of curvature of the curve $y = e^x$ at the point for which $x = 0$.

(2) Find the radius and the center of curvature of the curve
$$y = x\left(\frac{x}{2} - 1\right)$$ at the point for which $x = 2$.

(3) Find the point or points of curvature unity in the curve $y = x^2$.

(4) Find the radius and the center of curvature of the curve $xy = m$ at the point for which $x = \sqrt{m}$.

(5) Find the radius and the center of curvature of the curve $y^2 = 4ax$ at the point for which $x = 0$.

(6) Find the radius and the center of curvature of the curve $y = x^3$ at the points for which $x = \pm 0.9$ and also $x = 0$.

(7) Find the radius of curvature and the coordinates of the center of curvature of the curve

$$y = x^2 - x + 2$$

at the two points for which $x = 0$ and $x = 1$, respectively. Find also the maximum or minimum value of y. Verify graphically all your results.

(8) Find the radius of curvature and the coordinates of the center of curvature of the curve

$$y = x^3 - x - 1$$

at the points for which $x = -2$, $x = 0$, and $x = 1$.

(9) Find the coordinates of the point or points of inflection of the curve $y = x^3 + x^2 + 1$.

(10) Find the radius of curvature and the coordinates of the center of curvature of the curve

$$y = (4x - x^2 - 3)^{\frac{1}{2}}$$

at the points for which $x = 1.2$, $x = 2$ and $x = 2.5$. What is this curve?

(11) Find the radius and the center of curvature of the curve $y = x^3 - 3x^2 + 2x + 1$ at the points for which $x = 0$, $x = +1.5$. Find also the position of the point of inflection.

(12) Find the radius and center of curvature of the curve $y = \sin \theta$ at the points for which $\theta = \frac{\pi}{4}$ and $\theta = \frac{\pi}{2}$. Find the position of the points of inflection.

(13) Draw a circle of radius 3, the center of which has for its

coordinates $x = 1$, $y = 0$. Deduce the equation of such a circle from first principles. Find by calculation the radius of curvature and the coordinates of the center of curvature for several suitable points, as accurately as possible, and verify that you get the known values.

(14) Find the radius and center of curvature of the curve $y = \cos \theta$ at the points for which $\theta = 0$, $\theta = \dfrac{\pi}{4}$ and $\theta = \dfrac{\pi}{2}$.

(15) Find the radius of curvature and the center of curvature of the ellipse $\dfrac{x^2}{a^2} + \dfrac{y^2}{b^2} = 1$ at the points for which $x = 0$ and at the points for which $y = 0$.

(16) When a curve is defined by equations in the form

$$x = F(\theta), \quad y = f(\theta)$$

the radius r of curvature is given by

$$r = \left\{ \left(\frac{dx}{d\theta} \right)^2 + \left(\frac{dy}{d\theta} \right)^2 \right\}^{\frac{3}{2}} \bigg/ \left(\frac{dx}{d\theta} \cdot \frac{d^2y}{d\theta^2} - \frac{dy}{d\theta} \cdot \frac{d^2x}{d\theta^2} \right)$$

Apply the formula to find r for the curve

$$x = a(\theta - \sin \theta), \quad y = a(1 - \cos \theta)$$

Chapter XXIII

HOW TO FIND THE LENGTH OF AN ARC ON A CURVE

Since an arc on any curve is made up of a lot of little bits of straight lines joined end to end, if we could add all these little bits, we would get the length of the arc. But we have seen that to add a lot of little bits together is precisely what is called integration, so that it is likely that, since we know how to integrate, we can find also the length of an arc on any curve, provided that the equation of the curve is such that it lends itself to integration.

If *MN* is an arc on any curve, the length *s* of which is required (see Fig. 66*a*), if we call "a little bit" of the arc *ds*, then we see at once that

$$(ds)^2 = (dx)^2 + (dy)^2$$

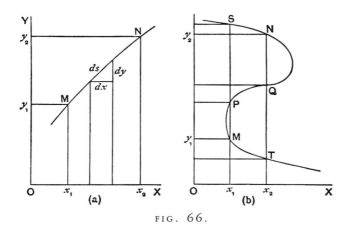

FIG. 66.

or either

$$ds = \sqrt{1 + \left(\frac{dx}{dy}\right)^2}\, dy \quad \text{or} \quad ds = \sqrt{1 + \left(\frac{dy}{dx}\right)^2}\, dx$$

Now the arc MN is made up of the sum of all the little bits ds between M and N, that is, between x_1 and x_2, or between y_1 and y_2, so that we get either

$$s = \int_{x_1}^{x_2} \sqrt{1 + \left(\frac{dy}{dx}\right)^2}\, dx \quad \text{or} \quad s = \int_{y_1}^{y_2} \sqrt{1 + \left(\frac{dx}{dy}\right)^2}\, dy$$

That is all!

The second integral is useful when there are several points of the curve corresponding to the given values of x (as in Fig. 66b). In this case the integral between x_1 and x_2 leaves a doubt as to the exact portion of the curve, the length of which is required. It may be ST, instead of MN, or SQ; by integrating between y_1 and y_2 the uncertainty is removed, and in this case one should use the second integral.

If instead of x and y coordinates—or Cartesian coordinates, as they are named from the French mathematician Descartes, who invented them—we have r and θ coordinates (or polar coordinates); then, if MN be a small arc of length ds on any curve, the length s of which is required (see Fig. 67), O being the pole, then the distance ON will generally differ from OM by a small amount dr. If the small angle MON is called $d\theta$, then, the polar coordinates of the point M being θ and r, those of N are $(\theta + d\theta)$ and $(r + dr)$. Let MP be perpendicular to ON, and let $OR = OM$; then

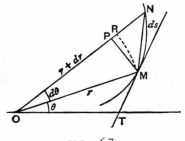

FIG. 67.

$RN = dr$, and this is very nearly the same as PN, as long as $d\theta$ is a very small angle. Also $RM = r\, d\theta$, and RM is very nearly equal to PM, and the arc MN is very nearly equal to the chord MN. In fact we can write $PN = dr$, $PM = r\, d\theta$, and arc MN = chord MN without appreciable error, so that we have:

$$(ds)^2 = (\text{chord } MN)^2 = \overline{PN}^2 + \overline{PM}^2 = dr^2 + r^2 d\theta^2$$

Dividing by $d\theta^2$ we get $\left(\dfrac{ds}{d\theta}\right)^2 = r^2 + \left(\dfrac{dr}{d\theta}\right)^2$; hence

$$\frac{ds}{d\theta} = \sqrt{r^2 + \left(\frac{dr}{d\theta}\right)^2} \quad \text{and} \quad ds = \sqrt{r^2 + \left(\frac{dr}{d\theta}\right)^2}\, d\theta$$

hence, since the length s is made up of the sum of all the little bits ds, between values of $\theta = \theta_1$ and $\theta = \theta_2$, we have

$$s = \int_{\theta_1}^{\theta_2} ds = \int_{\theta_1}^{\theta_2} \sqrt{r^2 + \left(\frac{dr}{d\theta}\right)^2}\, d\theta$$

We can proceed at once to work out a few examples.

Example 1.   The equation of a circle, the center of which is at the origin—or intersection of the axis of x with the axis of y—is $x^2 + y^2 = r^2$; find the length of an arc of one quadrant.

$$y^2 = r^2 - x^2 \text{ and } 2y\, dy = -2x\, dx, \text{ so that } \frac{dy}{dx} = -\frac{x}{y}$$

hence

$$s = \int \sqrt{\left[1 + \left(\frac{dy}{dx}\right)^2\right]}\, dx = \int \sqrt{\left(1 + \frac{x^2}{y^2}\right)}\, dx$$

and since $y^2 = r^2 - x^2$,

$$s = \int \sqrt{\left(1 + \frac{x^2}{r^2 - x^2}\right)}\, dx = \int \frac{r\, dx}{\sqrt{r^2 - x^2}}$$

The length we want—one quadrant—extends from a point for which $x = 0$ to another point for which $x = r$. We express this by writing

$$s = \int_{x=0}^{x=r} \frac{r\,dx}{\sqrt{r^2 - x^2}}$$

or, more simply, by writing

$$s = \int_{0}^{r} \frac{r\,dx}{\sqrt{r^2 - x^2}}$$

the 0 and r to the right of the sign of integration merely meaning that the integration is only to be performed on a portion of the curve, namely, that between $x = 0$, $x = r$, as we have seen.

Here is a fresh integral for you! Can you manage it?

In Chapter XV we differentiated $y = \arcsin x$ and found $\frac{dy}{dx} = \frac{1}{\sqrt{1 - x^2}}$. If you have tried all sorts of variations of the given examples (as you ought to have done!), you perhaps tried to differentiate something like $y = a \arcsin \frac{x}{a}$, which gave

$$\frac{dy}{dx} = \frac{a}{\sqrt{a^2 - x^2}} \quad \text{or} \quad dy = \frac{a\,dx}{\sqrt{a^2 - x^2}}$$

that is, just the same expression as the one we have to integrate here.

Hence $s = \int \frac{r\,dx}{\sqrt{r^2 - x^2}} = r \arcsin \frac{x}{r} + C$, C being a constant.

As the integration is only to be made between $x = 0$ and $x = r$, we write

$$s = \int_{0}^{r} \frac{r\,dx}{\sqrt{r^2 - x^2}} = \left[r \arcsin \frac{x}{r} + C \right]_{0}^{r}$$

proceeding then as explained in Example (1), Chapter XIX, we get

$$s = r \arcsin \frac{r}{r} + C - r \arcsin \frac{0}{r} - C, \quad \text{or} \quad s = r \times \frac{\pi}{2}$$

since arcsin 1 is 90° or $\dfrac{\pi}{2}$ and arc-

sin 0 is zero, and the constant C disappears, as has been shown.

The length of the quadrant is therefore $\dfrac{\pi r}{2}$, and the length of the circumference, being four times this, is $4 \times \dfrac{\pi r}{2} = 2\pi r$.

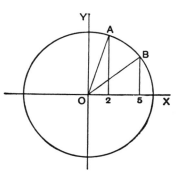

FIG. 68.

Example 2. Find the length of the arc AB between $x_1 = 2$ and $x_2 = 5$, in the circumference $x^2 + y^2 = 6^2$ (see Fig. 68).

Here, proceeding as in the previous example,

$$s = \left[r\,\text{arcsin}\,\frac{x}{r} + C \right]_{x_1}^{x_2} = \left[6\,\text{arcsin}\,\frac{x}{6} + C \right]_{2}^{5}$$

$$= 6\left[\text{arcsin}\,\frac{5}{6} - \text{arcsin}\,\frac{2}{6} \right] = 6(0.9851 - 0.3398)$$

$$= 3.8716 \text{ (the arcs being expressed in radians).}$$

It is always well to check results obtained by a new and yet unfamiliar method. This is easy, for

$$\cos AOX = \tfrac{2}{6} = \tfrac{1}{3} \text{ and } \cos BOX = \tfrac{5}{6}$$

hence $AOB = AOX - BOX = \text{arccos}\,\tfrac{1}{3} = \text{arccos}\,\tfrac{5}{6} = 0.6453$ radians, and the length is $6 \times 0.6453 = 3.8716$.

Example 3. Find the length of an arc of the curve

$$y = \frac{a}{2}\left\{ e^{\frac{x}{a}} + e^{-\frac{x}{a}} \right\}$$

between $x = 0$ and $x = a$. (This curve is the *catenary*.)

$$y = \frac{a}{2}e^{\frac{x}{a}} + \frac{a}{2}e^{-\frac{x}{a}}, \frac{dy}{dx} = \frac{1}{2}\left\{e^{\frac{x}{a}} - e^{-\frac{x}{a}}\right\}$$

$$s = \int \sqrt{1 + \frac{1}{4}\left\{e^{\frac{x}{a}} - e^{-\frac{x}{a}}\right\}^2} \, dx$$

$$= \frac{1}{2} \int \sqrt{4 + e^{\frac{2x}{a}} + e^{-\frac{2x}{a}} - 2e^{\frac{x}{a} - \frac{x}{a}}} \, dx$$

Now

$$e^{\frac{x}{a} - \frac{x}{a}} = e^0 = 1, \text{ so that } s = \frac{1}{2} \int \sqrt{2 + e^{\frac{2x}{a}} + e^{-\frac{2x}{a}}} \, dx$$

we can replace 2 by $2 \times e^0 = 2 \times e^{\frac{x}{a} - \frac{x}{a}}$; then

$$s = \frac{1}{2} \int \sqrt{e^{\frac{2x}{a}} + 2e^{\frac{x}{a} - \frac{x}{a}} + e^{-\frac{2x}{a}}} \, dx$$

$$= \frac{1}{2} \int \sqrt{\left(e^{\frac{x}{a}} + e^{-\frac{x}{a}}\right)^2} \, dx = \frac{1}{2} \int \left(e^{\frac{x}{a}} + e^{-\frac{x}{a}}\right) dx$$

$$= \frac{1}{2} \int e^{\frac{x}{a}} \, dx + \frac{1}{2} \int e^{-\frac{x}{a}} \, dx = \frac{a}{2} \left[e^{\frac{x}{a}} - e^{-\frac{x}{a}}\right]$$

Here $s = \frac{a}{2}\left[e^{\frac{x}{a}} - e^{-\frac{x}{a}}\right]_0^a = \frac{a}{2}[e^1 - e^{-1} - 1 + 1]$

and $s = \frac{a}{2}\left(e - \frac{1}{e}\right) = 1.1752a.$

Example 4.   A curve is such that the length of the tangent at any point P (see Fig. 69) from P to the intersection T of the tangent with a fixed line AB is a constant length a. Find an expression for

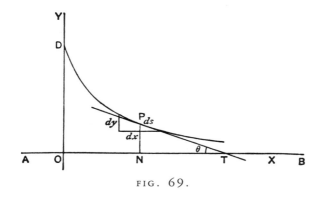

FIG. 69.

an arc of this curve—which is called the tractrix[1]—and find the length, when $a = 3$, between the ordinates $y = a$ and $y = 1$.

We shall take the fixed line for the axis of x. The point D, with $DO = a$, is a point on the curve, which must be tangent to OD at D. We take OD as the axis of y. $PT = a$, $PN = y$, $ON = x$.

If we consider a small portion ds of the curve, at P, then

$\sin \theta = \dfrac{dy}{ds} = -\dfrac{y}{a}$ (minus because the curve slopes *downwards* to the right.

Hence $\dfrac{ds}{dy} = -\dfrac{a}{y}$, $ds = -a\dfrac{dy}{y}$ and $s = -a\displaystyle\int\dfrac{dy}{y}$, that is,

$$s = -a \ln y + C$$

When $x = 0$, $s = 0$, $y = a$, so that $0 = -a \ln a + C$, and $C = a \ln a$.

It follows that $s = a \ln a - a \ln y = a \ln \dfrac{a}{y}$.

1. The tractrix is the involute of the catenary. The involute of a curve is the curve traced by the end of a taut thread as it is unwound from a given curve.

The origin of the name tractrix is interesting. It can be demonstrated by the following experiment. Attach a length of string to a small object. Place the object at the center of a large rectangular table, then extend the string horizontally until it hangs over, say, the table's right edge. Grab the string where it touches the table edge, then move it along the edge. The track generated by the object as it is dragged along the table is a tractrix.—M.G.

When $a = 3$, s between $y = a$ and $y = 1$ is therefore

$$s = 3\left[\ln\frac{3}{y}\right]_1^3 = 3(\ln 1 - \ln 3) = 3 \times (0 - 1.0986)$$

$$= -3.296 \text{ or } 3.296,$$

as the sign − refers merely to the direction in which the length was measured, from D to P, or from P to D.

Note that this result has been obtained without a knowledge of the equation of the curve. This is sometimes possible. In order to get the length of an arc between two points given by their abscissae, however, it is necessary to know the equation of the curve; this is easily obtained as follows:

$$\frac{dy}{dx} = -\tan\theta = -\frac{y}{\sqrt{a^2 - y^2}}, \text{ since } PT = a$$

hence $dx = -\dfrac{\sqrt{a^2 - y^2}\, dy}{y}$, and $x = -\displaystyle\int \dfrac{\sqrt{a^2 - y^2}\, dy}{y}$

The integration will give us a relation between x and y, which is the equation of the curve.

To effect the integration, let $u^2 = a^2 - y^2$, then

$$2u\, du = -2y\, dy, \quad \text{or} \quad u\, du = -y\, dy$$

$$x = \int \frac{u^2\, du}{y^2} = \int \frac{u^2\, du}{a^2 - u^2} = \int \frac{a^2 - (a^2 - u^2)}{a^2 - u^2} \cdot du$$

$$= a^2 \int \frac{du}{a^2 - u^2} - \int du$$

$$= a^2 \cdot \frac{1}{2a} \ln \frac{a + u}{a - u} - u + C$$

$$= \tfrac{1}{2}a \ln \frac{(a + u)(a + u)}{(a - u)(a + u)} - u + C$$

$$= a \ln \frac{a + u}{\sqrt{a^2 - u^2}} - u + C$$

We have then, finally,

$$x = a \ln \frac{a + \sqrt{a^2 - y^2}}{y} - \sqrt{a^2 - y^2} + C$$

When $x = 0$, $y = a$, so that $0 = a \ln 1 - 0 + C$, and $C = 0$; the equation of the tractrix is therefore

$$x = a \ln \frac{a + \sqrt{a^2 - y^2}}{y} - \sqrt{a^2 - y^2}$$

Example 5.   Find the length of an arc of the logarithmic spiral $r = e^{\theta}$ between $\theta = 0$ and $\theta = 1$ radian.

Do you remember differentiating $y = e^x$? It is an easy one to remember, for it remains always the same whatever is done to it: $\frac{dy}{dx} = e^x$.

Here, since $\qquad r = e^{\theta}, \dfrac{dr}{d\theta} = e^{\theta} = r$

If we reverse the process and integrate $\int e^{\theta} \, d\theta$ we get back to $r + C$, the constant C being always introduced by such a process, as we have seen in Chap. XVII.

It follows that

$$s = \int \sqrt{\left[r^2 + \left(\frac{dr}{d\theta} \right)^2 \right]} \, d\theta = \int \sqrt{(r^2 + r^2)} \, d\theta$$

$$= \sqrt{2} \int r \, d\theta = \sqrt{2} \int e^{\theta} \, d\theta = \sqrt{2} \, (e^{\theta} + C)$$

Integrating between the two given values $\theta = 0$ and $\theta = 1$, we get

$$s = \int_0^1 \sqrt{\left[r^2 + \left(\frac{dr}{d\theta} \right)^2 \right]} \, d\theta = \left[\sqrt{2(e^{\theta} + C)} \right]_0^1$$

$$= \sqrt{2} e^1 - \sqrt{2} e^0 = \sqrt{2}(e - 1)$$

$$= 1.41 \times 1.718 = 2.43$$

Example 6.   Find the length of an arc of the logarithmic spiral $r = e^\theta$ between $\theta = 0$ and $\theta = \theta_1$.

As we have just seen,

$$s = \sqrt{2} \int_0^{\theta_1} e^\theta \, d\theta = \sqrt{2}[e^{\theta_1} - e^0] = \sqrt{2}(e^{\theta_1} - 1)$$

Example 7.   As a last example let us work fully a case leading to a typical integration which will be found useful for several of the exercises found at the end of this chapter. Let us find the expression for the length of an arc of the curve $y = \dfrac{a}{2}x^2 + 3$.

$$\frac{dy}{dx} = ax, \quad s = \int \sqrt{1 + a^2x^2} \, dx$$

To work out this integral, let $ax = \sinh z$, then $a \, dx = \cosh z \, dz$, and $1 + a^2x^2 = 1 + \sinh^2 z = \cosh^2 z$;

$$s = \frac{1}{a} \int \cosh^2 z \, dz = \frac{1}{4a} \int (e^{2z} + 2 + e^{-2z}) dz$$

$$= \frac{1}{4a}[\tfrac{1}{2}e^{2z} + 2z - \tfrac{1}{2}e^{-2z}] = \frac{1}{8a}[(e^z)^2 - (e^{-z})^2 + 4z]$$

$$= \frac{1}{8a}(e^z - e^{-z})(e^z + e^{-z}) + \frac{z}{2a}$$

$$= \frac{1}{2a}(\sinh z \cosh z + z) = \frac{1}{2a}\left(ax\sqrt{1 + a^2x^2} + z\right)$$

To turn z back into terms of x, we have

$$ax = \sinh z = \tfrac{1}{2}(e^z - e^{-z})$$

Multiply out by $2e^z$,

$$2axe^z = e^{2z} - 1$$

or $$(e^z)^2 - 2ax(e^z) - 1 = 0$$

This is a quadratic equation in e^z, and taking the positive root:

$$e^z = \tfrac{1}{2}\left(2ax + \sqrt{4a^2x^2 + 4}\right) = ax + \sqrt{1 + a^2x^2}$$

Taking natural logarithms:

$$z = \ln\left(ax + \sqrt{1 + a^2x^2}\right)$$

Hence, the integral becomes finally:

$$s = \int \sqrt{1 + a^2x^2}\,dx = \frac{x}{2}\sqrt{1 + a^2x^2} + \frac{1}{2a}\ln\left(ax + \sqrt{1 + a^2x^2}\right)$$

From several of the foregoing examples, some very important integrals and relations have been worked out. As these are of great use in solving many other problems, it will be an advantage to collect them here for future reference.

Inverse Hyperbolic Functions

If $x = \sinh z$, z is written inversely as $\sinh^{-1}x$;

and
$$z = \sinh^{-1}x = \ln\left(x + \sqrt{x^2 + 1}\right).$$

Similarly, if $x = \cosh z$,

$$z = \cosh^{-1}x = \ln\left(x + \sqrt{x^2 - 1}\right)$$

Irrational Quadratic Integrals

(i) $\displaystyle\int \frac{\sqrt{a^2 - x^2}}{x}\,dx = \sqrt{a^2 - x^2} - a\ln\frac{a + \sqrt{a^2 - x^2}}{x} + C$

(ii) $\displaystyle\int \sqrt{a^2 + x^2}\,dx = \tfrac{1}{2}x\sqrt{a^2 + x^2} + \tfrac{1}{2}a^2\ln\left(x + \sqrt{a^2 + x^2}\right) + C$

To these may be added:

(iii) $\displaystyle\int \frac{dx}{\sqrt{a^2 + x^2}} = \ln\left(x + \sqrt{a^2 + x^2}\right) + C$

For, if $x = a \sinh u$, $dx = a \cosh u \, du$, and

$$\int \frac{dx}{\sqrt{a^2 + x^2}} = \int du = u + C' = \sinh^{-1} \frac{x}{a} + C'$$

$$= \ln \frac{x + \sqrt{a^2 + x^2}}{a} + C'$$

$$= \ln \left(x + \sqrt{a^2 + x^2} \right) + C$$

You ought now to be able to attempt with success the follow-ing exercises. You will find it interesting as well as instructive to plot the curves and verify your results by measurement where possible.

EXERCISE XXII

(1) Find the length of the line $y = 3x + 2$ between the two points for which $x = 1$ and $x = 4$.

(2) Find the length of the line $y = ax + b$ between the two points for which $x = -1$ and $x = a^2$.

(3) Find the length of the curve $y = \frac{2}{3} x^{\frac{3}{2}}$ between the two points for which $x = 0$ and $x = 1$.

(4) Find the length of the curve $y = x^2$ between the two points for which $x = 0$ and $x = 2$.

(5) Find the length of the curve $y = mx^2$ between the two points for which $x = 0$ and $x = \dfrac{1}{2m}$.

(6) Find the length of the curve $x = a \cos \theta$ and $y = a \sin \theta$ be-tween $\theta = \theta_1$ and $\theta = \theta_2$.

(7) Find the length of the arc of the curve $r = a \sec \theta$ from $\theta = 0$ to an arbitrary point on the curve.

(8) Find the length of the arc of the curve $y^2 = 4ax$ between $x = 0$ and $x = a$.

(9) Find the length of the arc of the curve $y = x \left(\dfrac{x}{2} - 1 \right)$ be-tween $x = 0$ and $x = 4$.

(10) Find the length of the arc of the curve $y = e^x$ between $x = 0$ and $x = 1$.

(*Note.* This curve is in rectangular coordinates, and is not the same curve as the logarithmic spiral $r = e^{\theta}$ which is in polar coordinates. The two equations are similar, but the curves are quite different.)

(11) A curve is such that the coordinates of a point on it are $x = a(\theta - \sin \theta)$ and $y = a(1 - \cos \theta)$, θ being a certain angle which varies between 0 and 2π. Find the length of the curve. (It is called a *cycloid*.)[2]

(12) Find the length of an arc of the curve $y = \ln \sec x$ between $x = 0$ and $x = \dfrac{\pi}{4}$ radians.

(13) Find the expression for the length of an arc of the curve $y^2 = \dfrac{x^3}{a}$.

(14) Find the length of the curve $y^2 = 8x^3$ between the two points for which $x = 1$ and $x = 2$.

(15) Find the length of the curve $y^{\frac{2}{3}} + x^{\frac{2}{3}} = a^{\frac{2}{3}}$ between $x = 0$ and $x = a$.

(16) Find the length of the curve $r = a(1 - \cos \theta)$ between $\theta = 0$ and $\theta = \pi$.

You have now been personally conducted over the frontiers into the enchanted land. And in order that you may have a handy reference to the principal results, the author, in bidding you farewell, begs to present you with a passport in the shape of a convenient collection of standard forms. In the middle column are set down a number of the functions which most commonly occur. The results of differentiating them are set down on the left; the results of integrating them are set down on the right. May you find them useful!

2. See the paragraph on the cycloid in this book's appendix.—M.G.

Table of Standard Forms

$\dfrac{dy}{dx}$	y	$\displaystyle\int y\, dx$
	Algebraic	
1	x	$\frac{1}{2}x^2 + C$
0	a	$ax + C$
1	$x \pm a$	$\frac{1}{2}x^2 \pm ax + C$
a	ax	$\frac{1}{2}ax^2 + C$
$2x$	x^2	$\frac{1}{3}x^3 + C$
nx^{n-1}	x^n	$\dfrac{1}{n+1}x^{n+1} + C$
$-x^{-2}$	x^{-1}	$\ln x + C$
$\dfrac{du}{dx} \pm \dfrac{dv}{dx} \pm \dfrac{dw}{dx}$	$u \pm v \pm w$	$\displaystyle\int u\, dx \pm \int v\, dx \pm \int w\, dx$
$u\dfrac{dv}{dx} + v\dfrac{du}{dx}$	uv	Put $v = \dfrac{dy}{dx}$ and integrate by parts
$\dfrac{v\dfrac{du}{dx} - u\dfrac{dv}{dx}}{v^2}$	$\dfrac{u}{v}$	No general form known
$\dfrac{du}{dx}$	u	$\displaystyle\int u\, dx = ux - \int x\, du + C$
	Exponential and Logarithmic	
e^x	e^x	$e^x + C$
x^{-1}	$\ln x$	$x(\ln x - 1) + C$
$0.4343x^{-1}$	$\log_{10} x$	$0.4343x(\ln x - 1) + C$
$a^x \ln a$	a^x	$\dfrac{a^x}{\ln a} + C$
	Trigonometrical	
$\cos x$	$\sin x$	$-\cos x + C$
$-\sin x$	$\cos x$	$\sin x + C$
$\sec^2 x$	$\tan x$	$-\ln \cos x + C$

Circular (Inverse)

$\dfrac{1}{\sqrt{1-x^2}}$	arcsin x	$x \arcsin x + \sqrt{1-x^2} + C$
$-\dfrac{1}{\sqrt{1-x^2}}$	arccos x	$x \arccos x - \sqrt{1-x^2} + C$
$\dfrac{1}{1+x^2}$	arctan x	$x \arctan x - \frac{1}{2}\ln(1+x^2) + C$

Hyperbolic

cosh x	sinh x	cosh x + C
sinh x	cosh x	sinh x + C
$\operatorname{sech}^2 x$	tanh x	ln cosh x + C

Miscellaneous

$-\dfrac{1}{(x+a)^2}$	$\dfrac{1}{x+a}$	$\ln|x+a| + C$
$-\dfrac{x}{(a^2+x^2)^{\frac{3}{2}}}$	$\dfrac{1}{\sqrt{a^2+x^2}}$	$\ln\left(x+\sqrt{a^2+x^2}\right) + C$
$\pm\dfrac{b}{(a\pm bx)^2}$	$\dfrac{1}{a\pm bx}$	$\pm\dfrac{1}{b}\ln|a\pm bx| + C$
$\dfrac{-3a^2 x}{(a^2+x^2)^{\frac{5}{2}}}$	$\dfrac{a^2}{(a^2+x^2)^{\frac{3}{2}}}$	$\dfrac{x}{\sqrt{a^2+x^2}} + C$
$a \cos ax$	$\sin ax$	$-\dfrac{1}{a}\cos ax + C$
$-a \sin ax$	$\cos ax$	$\dfrac{1}{a}\sin ax + C$
$a \sec^2 ax$	$\tan ax$	$-\dfrac{1}{a}\ln|\cos ax| + C$
$\sin 2x$	$\sin^2 x$	$\dfrac{x}{2} - \dfrac{\sin 2x}{4} + C$
$-\sin 2x$	$\cos^2 x$	$\dfrac{x}{2} + \dfrac{\sin 2x}{4} + C$

Miscellaneous

$n \cdot \sin^{n-1}x \cdot \cos x$	$\sin^n x$	$-\dfrac{\cos x}{n} \sin^{n-1}x$ $+\dfrac{n-1}{n} \displaystyle\int \sin^{n-2}x \, dx + C$
$-\dfrac{\cos x}{\sin^2 x}$	$\dfrac{1}{\sin x}$	$\ln \left| \tan \dfrac{x}{2} \right| + C$
$-\dfrac{\sin 2x}{\sin^4 x}$	$\dfrac{1}{\sin^2 x}$	$-\cot x + C$
$\dfrac{\sin^2 x - \cos^2 x}{\sin^2 x \cdot \cos^2 x}$	$\dfrac{1}{\sin x \cdot \cos x}$	$\ln \,|\tan x| + C$
$n \cdot \sin mx \cdot \cos nx +$ $m \cdot \sin nx \cdot \cos mx$	$\sin mx \cdot \sin nx$	$\dfrac{\sin (m - n)x}{2(m - n)} -$ $\dfrac{\sin (m + n)x}{2 \,(m + n)} + C$
$a \sin 2ax$	$\sin^2 ax$	$\dfrac{x}{2} - \dfrac{\sin 2ax}{4a} + C$
$-a \sin 2ax$	$\cos^2 ax$	$\dfrac{x}{2} + \dfrac{\sin 2ax}{4a} + C$

EPILOGUE AND APOLOGUE

———∞∞∞———

It may be confidently assumed that when this book *Calculus Made Easy* falls into the hands of the professional mathematicians, they will (if not too lazy) rise up as one man, and damn it as being a thoroughly bad book. Of that there can be, from their point of view, no possible manner of doubt whatever. It commits several most grievous and deplorable errors.

First, it shows how ridiculously easy most of the operations of the calculus really are.

Secondly, it gives away so many trade secrets. By showing you that *what one fool can do, other fools can do also,* it lets you see that these mathematical swells, who pride themselves on having mastered such an awfully difficult subject as the calculus, have no such great reason to be puffed up. They like you to think how terribly difficult it is, and don't want that superstition to be rudely dissipated.

Thirdly, among the dreadful things they will say about "So Easy" is this: that there is an utter failure on the part of the author to demonstrate with rigid and satisfactory completeness the validity of sundry methods which he has presented in simple fashion, and has even *dared to use* in solving problems! But why should he not? You don't forbid the use of a watch to every person who does not know how to make one? You don't object to the musician playing on a violin that he has not himself constructed. You don't teach the rules of syntax to children until they have already become fluent in the *use* of speech. It would be equally absurd to require general rigid demonstrations to be expounded to beginners in the calculus.

One other thing will the professed mathematicians say about this thoroughly bad and vicious book: that the reason why it is *so easy* is because the author has left out all the things that are really difficult. And the ghastly fact about this accusation is that—*it is true!* That is, indeed, why the book has been written—written for the legion of innocents who have hitherto been deterred from acquiring the elements of the calculus by the stupid way in which its teaching is almost always presented. Any subject can be made repulsive by presenting it bristling with difficulties. The aim of this book is to enable beginners to learn its language, to acquire familiarity with its endearing simplicities, and to grasp its powerful methods of solving problems, without being compelled to toil through the intricate out-of-the-way (and mostly irrelevant) mathematical gymnastics so dear to the unpractical mathematician.

There are amongst young engineers a number on whose ears the adage that *what one fool can do, another can,* may fall with a familiar sound. They are earnestly requested not to give the author away, nor to tell the mathematicians what a fool he really is.

ANSWERS
Exercises I

(1) $\dfrac{dy}{dx} = 13x^{12}$　　(2) $\dfrac{dy}{dx} = -\dfrac{3}{2}x^{-\frac{5}{2}}$　　(3) $\dfrac{dy}{dx} = 2ax^{2a-1}$

(4) $\dfrac{du}{dt} = 2.4t^{1.4}$　　(5) $\dfrac{dz}{du} = \dfrac{1}{3}u^{-\frac{2}{3}}$　　(6) $\dfrac{dy}{dx} = -\dfrac{5}{3}x^{-\frac{8}{3}}$

(7) $\dfrac{du}{dx} = -\dfrac{8}{5}x^{-\frac{13}{5}}$　　　　(8) $\dfrac{dy}{dx} = 2ax^{a-1}$

(9) $\dfrac{dy}{dx} = \dfrac{3}{q}x^{\frac{3-q}{q}}$　　　　(10) $\dfrac{dy}{dx} = -\dfrac{m}{n}x^{-\frac{m+n}{n}}$

Exercises II

(1) $\dfrac{dy}{dx} = 3ax^2$　　(2) $\dfrac{dy}{dx} = 13 \times \tfrac{3}{2}x^{\frac{1}{2}}$　　(3) $\dfrac{dy}{dx} = 6x^{-\frac{1}{2}}$

(4) $\dfrac{dy}{dx} = \dfrac{1}{2}c^{\frac{1}{2}}x^{-\frac{1}{2}}$　　(5) $\dfrac{du}{dz} = \dfrac{an}{c}z^{n-1}$　　(6) $\dfrac{dy}{dt} = 2.36t$

(7) $\dfrac{dl_t}{dt} = 0.000012 \times l_0$

(8) $\dfrac{dc}{dV} = abV^{b-1}$, 0.98, 3.00 and 7.46 candle power per volt respectively.

(9) $\dfrac{dn}{dD} = -\dfrac{1}{LD^2}\sqrt{\dfrac{gT}{\pi\sigma}}$, $\dfrac{dn}{dL} = -\dfrac{1}{DL^2}\sqrt{\dfrac{gT}{\pi\sigma}}$

$\dfrac{dn}{d\sigma} = -\dfrac{1}{2DL}\sqrt{\dfrac{gT}{\pi\sigma^3}}$, $\dfrac{dn}{dT} = \dfrac{1}{2DL}\sqrt{\dfrac{g}{\pi\sigma T}}$

(10) $\dfrac{\text{Rate of change of } P \text{ when } t \text{ varies}}{\text{Rate of change of } P \text{ when } D \text{ varies}} = -\dfrac{D}{t}$

(11) 2π, $2\pi r$, πl, $\tfrac{2}{3}\pi rh$, $8\pi r$, $4\pi r^2$.

Exercises III

(1) (a) $1 + x + \dfrac{x^2}{2} + \dfrac{x^3}{6} + \dfrac{x^4}{24} + \ldots$ (b) $2ax + b$ (c) $2x + 2a$

(d) $3x^2 + 6ax + 3a^2$

(2) $\dfrac{dw}{dt} = a - bt$ (3) $\dfrac{dy}{dx} = 2x$

(4) $14110x^4 - 65404x^3 - 2244x^2 + 8192x + 1379$

(5) $\dfrac{dx}{dy} = 2y + 8$ (6) $185.9022654x^2 + 154.36334$

(7) $\dfrac{-5}{(3x + 2)^2}$ (8) $\dfrac{6x^4 + 6x^3 + 9x^2}{(1 + x + 2x^2)^2}$

(9) $\dfrac{ad - bc}{(cx + d)^2}$ (10) $\dfrac{anx^{-n-1} + bnx^{n-1} + 2nx^{-1}}{(x^{-n} + b)^2}$

(11) $b + 2ct$

(12) $R_0(a + 2bt),\ R_0\left(a + \dfrac{b}{2\sqrt{t}}\right),\ -\dfrac{R_0(a + 2bt)}{(1 + at + bt^2)^2}$ or $-\dfrac{R^2(a + 2bt)}{R_0{}^2}$

(13) $1.4340(0.000014t - 0.001024)$, -0.00117, -0.00107, -0.00097

(14) (a) $\dfrac{dE}{dl} = b + \dfrac{k}{i}$,   (b) $\dfrac{dE}{di} = -\dfrac{c + kl}{i^2}$

Exercises IV

(1) $17 + 24x$; 24 (2) $\dfrac{x^2 + 2ax - a}{(x + a)^2}$; $\dfrac{2a(a + 1)}{(x + a)^3}$

(3) $1 + x + \dfrac{x^2}{1 \times 2} + \dfrac{x^3}{1 \times 2 \times 3}$; $1 + x + \dfrac{x^2}{1 \times 2}$

(4) *(Exercises III):*

(1) (a) $\dfrac{d^2u}{dx^2} = \dfrac{d^3u}{dx^3} = 1 + x + \frac{1}{2}x^2 + \frac{1}{6}x^3 + \ldots$

(b) $2a$, 0   (c) 2, 0   (d) $6x + 6a$, 6

(2) $-b$, 0   (3) 2, 0

(4) $56440x^3 - 196212x^2 - 4488x + 8192.$
$169320x^2 - 392424x - 4488$

(5) 2, 0   (6) $371.80453x$, 371.80453

(7) $\dfrac{30}{(3x+2)^3}$, $-\dfrac{270}{(3x+2)^4}$

(Examples):

(1) $\dfrac{6a}{b^2}x$, $\dfrac{6a}{b^2}$ (2) $\dfrac{3a\sqrt{b}}{2\sqrt{x}} - \dfrac{6b\sqrt[3]{a}}{x^3}$, $\dfrac{18b\sqrt[3]{a}}{x^4} - \dfrac{3a\sqrt{b}}{4\sqrt{x^3}}$

(3) $\dfrac{2}{\sqrt[3]{\theta^8}} - \dfrac{1.056}{\sqrt[5]{\theta^{11}}}$, $\dfrac{2.3232}{\sqrt[5]{\theta^{16}}} - \dfrac{16}{3\sqrt[3]{\theta^{11}}}$

(4) $810t^4 - 648t^3 + 479.52t^2 - 139.968t + 26.64$
$3240t^3 - 1944t^2 + 959.04t - 139.968$

(5) $12x + 2$, 12   (6) $6x^2 - 9x$, $12x - 9$

(7) $\dfrac{3}{4}\left(\dfrac{1}{\sqrt{\theta}} + \dfrac{1}{\sqrt{\theta^5}} \right) + \dfrac{1}{4}\left(\dfrac{15}{\sqrt{\theta^7}} - \dfrac{1}{\sqrt{\theta^3}} \right)$

$\cdot \dfrac{3}{8}\left(\dfrac{1}{\sqrt{\theta^5}} - \dfrac{1}{\sqrt{\theta^3}} \right) - \dfrac{15}{8}\left(\dfrac{7}{\sqrt{\theta^9}} + \dfrac{1}{\sqrt{\theta^7}} \right)$

Exercises V

(2) 64; 147.2; and 0.32 feet per second

(3) $\dot{x} = a - gt$; $\ddot{x} = -g$ (4) 45.1 feet per second

(5) 12.4 feet per second per second. Yes.

(6) Angular velocity = 11.2 radians per second;
angular acceleration = 9.6 radians per second per second.

(7) $v = 20.4t^2 - 10.8, \quad a = 40.8t \quad$ 172.8 in./sec., 122.4 in./sec.^{2}

(8) $v = \dfrac{1}{30\sqrt[3]{(t-125)^2}}, a = -\dfrac{1}{45\sqrt[3]{(t-125)^5}}$

(9) $v = 0.8 - \dfrac{8t}{(4+t^2)^2}, a = \dfrac{24t^2 - 32}{(4+t^2)^3}$, 0.7926 and 0.00211

(10) $n = 2, n = 11$

Exercises VI

(1) $\dfrac{x}{\sqrt{x^2+1}}$

(2) $\dfrac{x}{\sqrt{x^2+a^2}}$

(3) $-\dfrac{1}{2\sqrt{(a+x)^3}}$

(4) $\dfrac{ax}{\sqrt{(a-x^2)^3}}$

(5) $\dfrac{2a^2 - x^2}{x^3\sqrt{x^2-a^2}}$

(6) $\dfrac{\frac{3}{2}x^2\left[\frac{8}{9}x(x^3+a) - (x^4+a)\right]}{(x^4+a)^{\frac{2}{3}}(x^3+a)^{\frac{3}{2}}}$

(7) $\dfrac{2a(x-a)}{(x+a)^3}$

(8) $\frac{5}{2}y^3$

(9) $\dfrac{1}{(1-\theta)\sqrt{1-\theta^2}}$

Exercises VII

(1) $\dfrac{dw}{dx} = -\dfrac{3x^2(3+3x^3)}{27\left(\frac{1}{2}x^3 + \frac{1}{4}x^6\right)^3}$

(2) $\dfrac{dv}{dx} = -\dfrac{12x}{\sqrt{1+\sqrt{2+3x^2}}\left(\sqrt{3}+4\sqrt{1+\sqrt{2+3x^2}}\right)^2}$

(3) $\dfrac{du}{dx} = -\dfrac{x^2\left(\sqrt{3}+x^3\right)}{\sqrt{\left[1+\left(1+\dfrac{x^3}{\sqrt{3}}\right)^2\right]^3}}$

(5) $\dfrac{dx}{d\theta} = a(1 - \cos\,\theta) = 2a\,\sin^2\tfrac{1}{2}\theta$

$\dfrac{dy}{d\theta} = a\,\sin\,\theta = 2a\,\sin\tfrac{1}{2}\theta\,\cos\tfrac{1}{2}\theta;\ \dfrac{dy}{dx} = \cot\tfrac{1}{2}\theta$

(6) $\dfrac{dx}{d\theta} = -3a\,\cos^2\,\theta\,\sin\,\theta,\ \dfrac{dy}{d\theta} = 3a\,\sin^2\,\theta\,\cos\,\theta;$

$\dfrac{dy}{dx} = -\tan\,\theta$

(7) $\dfrac{dy}{dx} = 2x\,\cot\,(x^2 - a^2)$

(8) Write $y = u - x$; find $\dfrac{dx}{du}, \dfrac{dy}{du}$, and then $\dfrac{dy}{dx}$

Exercises VIII

(2) 1.44

(4) $\dfrac{dy}{dx} = 3x^2 + 3$; and the numerical values are: $3, 3\tfrac{3}{4}, 6$, and 15.

(5) $\pm\sqrt{2}$

(6) $\dfrac{dy}{dx} = -\dfrac{4}{9}\dfrac{x}{y}$. Slope is zero where $x = 0$; and is $\pm\dfrac{1}{3\sqrt{2}}$ where $x = 1$.

(7) $m = 4, n = -3$

(8) Intersections at $x = 1, x = -3$. Angles $153°\,26', 2°\,28'$

(9) Intersection at $x = y = \tfrac{25}{7}$. Angle $16°\,16'$

(10) $x = \tfrac{1}{3}, y = 2\tfrac{1}{3}, b = -\tfrac{5}{3}$

Exercises IX

(1) Min.: $x = 0, y = 0$; max.: $x = -2, y = -4$

(2) $x = a$ $\qquad\qquad\qquad$ (4) $25\sqrt{3}$ square inches.

(5) $\dfrac{dy}{dx} = -\dfrac{10}{x^2} + \dfrac{10}{(8-x)^2}$; $x = 4$; $y = 5$

(6) Max. for $x = -1$; min. for $x = 1$.

(7) Join the middle points of the four sides.

(8) $r = \frac{2}{3}R$, $r = \dfrac{R}{2}$, no max.

(9) $r = R\sqrt{\dfrac{2}{3}}$, $r = \dfrac{R}{\sqrt{2}}$, $r = 0.8507R$

(10) At the rate of $8\sqrt{\pi}$ square feet per second.

(11) $r = \dfrac{R\sqrt{8}}{3}$

Exercises X

(1) Max.: $x = -2.19$, $y = 24.19$; min.: $x = 1.52$, $y = -1.38$

(2) $\dfrac{dy}{dx} = \dfrac{b}{a} - 2cx$; $\dfrac{d^2y}{dx^2} = -2c$; $x = \dfrac{b}{2ac}$ (a *maximum*)

(3) (*a*) One maximum and two minima.
   (*b*) One maximum.   ($x = 0$; other points unreal.)

(4) Min.: $x = 1.71$, $y = 6.13$ (5) Max.: $x = -.5$, $y = 4$

(6) Max.: $x = 1.414$, $y = 1.7678$. Min.: $x = -1.414$,
   $y = -1.7678$

(7) Max.: $x = -3.565$, $y = 2.12$. Min.: $x = +3.565$, $y = 7.88$

(8) $0.4N$, $0.6N$ (9) $x = \sqrt{\dfrac{a}{c}}$

(10) Speed 8.66 nautical miles per hour. Time taken 115.44 hours, total cost is \$2,251.11.

(11) Max. and min. for $x = 7.5$, $y = \pm 5.413$.

(12) Min.: $x = \frac{1}{2}$, $y = 0.25$; max.: $x = -\frac{1}{3}$, $y = 1.407$

Exercises XI

(1) $\dfrac{2}{x-3} + \dfrac{1}{x+4}$ (2) $\dfrac{1}{x-1} + \dfrac{2}{x-2}$ (3) $\dfrac{2}{x-3} + \dfrac{1}{x+4}$

(4) $\dfrac{5}{x-4} - \dfrac{4}{x-3}$ (5) $\dfrac{19}{13(2x+3)} - \dfrac{22}{13(3x-2)}$

(6) $\dfrac{2}{x-2} + \dfrac{4}{x-3} - \dfrac{5}{x-4}$

(7) $\dfrac{1}{6(x-1)} + \dfrac{11}{15(x+2)} + \dfrac{1}{10(x-3)}$

(8) $\dfrac{7}{9(3x+1)} + \dfrac{71}{63(3x-2)} - \dfrac{5}{7(2x+1)}$

(9) $\dfrac{1}{3(x-1)} + \dfrac{2x+1}{3(x^2+x+1)}$

(10) $x + \dfrac{2}{3(x+1)} + \dfrac{1-2x}{3(x^2-x+1)}$

(11) $\dfrac{3}{x+1} + \dfrac{2x+1}{x^2+x+1}$ (12) $\dfrac{1}{x-1} - \dfrac{1}{x-2} + \dfrac{2}{(x-2)^2}$

(13) $\dfrac{1}{4(x-1)} - \dfrac{1}{4(x+1)} + \dfrac{1}{2(x+1)^2}$

(14) $\dfrac{4}{9(x-1)} - \dfrac{4}{9(x+2)} - \dfrac{1}{3(x+2)^2}$

(15) $\dfrac{1}{x+2} - \dfrac{x-1}{x^2+x+1} - \dfrac{1}{(x^2+x+1)^2}$

(16) $\dfrac{5}{x+4} - \dfrac{32}{(x+4)^2} + \dfrac{36}{(x+4)^3}$

(17) $\dfrac{7}{9(3x-2)^2} + \dfrac{55}{9(3x-2)^3} + \dfrac{73}{9(3x-2)^4}$

(18) $\dfrac{1}{6(x-2)} + \dfrac{1}{3(x-2)^2} - \dfrac{x}{6(x^2+2x+4)}$

Exercises XII

(1) $ab(e^{ax} + e^{-ax})$ (2) $2at + \dfrac{2}{t}$ (3) $\ln n$ (5) npv^{n-1}

(6) $\dfrac{n}{x}$ (7) $\dfrac{3e^{-\frac{x}{x-1}}}{(x-1)^2}$ (8) $6xe^{-5x} - 5(3x^2 + 1)e^{-5x}$

(9) $\dfrac{ax^{a-1}}{x^a + a}$ (10) $\dfrac{15x^2 + 12x\sqrt{x} - 1}{2\sqrt{x}}$ (11) $\dfrac{1 - \ln(x+3)}{(x+3)^2}$

(12) $a^x(ax^{a-1} + x^a \ln a)$ (14) Min.: $y = 0.7$ for $x = 0.693$

(15) $\dfrac{1+x}{x}$ (16) $\dfrac{3}{x}(\ln ax)^2$

Exercises XIII

(2) $T = 34.625$; 159.45 minutes

(5) (a) $x^x(1 + \ln x)$;   (b) $2x\,(e^x)^x$;   (c) $e^{x^x} \times x^x\,(1 + \ln x)$

(6) 0.14 second   (7) (a) 1.642;   (b) 15.58

(8) $\mu = 0.00037$, 31.06 min

(9) i is 63.4% of i_0, 221.56 kilometers

(10) $k = 0.1339, 0.1445, 0.1555$, mean $= 0.1446$; percentage errors:—10.2%, 0.18% nil, +71.8%.

(11) Min. for $x = \dfrac{1}{e}$ (12) Max. for $x = e$

(13) Min. for $x = \ln a$

Exercises XIV

(1)   (i) $\dfrac{dy}{d\theta} = A \cos\left(\theta - \dfrac{\pi}{2}\right)$

     (ii) $\dfrac{dy}{d\theta} = 2 \sin\theta \cos\theta = \sin 2\theta$ and $\dfrac{dy}{d\theta} = 2\cos 2\theta$

(iii) $\dfrac{dy}{d\theta} = 3 \sin^2\theta \cos \theta$ and $\dfrac{dy}{d\theta} = 3 \cos 3\theta$

(2) $\theta = 45°$ or $\dfrac{\pi}{4}$ radians

(3) $\dfrac{dy}{dt} = -n \sin 2\pi nt$

(4) $a^x \, 1_n \, a \cos a^x$

(5) $\dfrac{-\sin x}{\cos x} = -\tan x$

(6) $18.2 \cos (x + 26°)$

(7) The slope is $\dfrac{dy}{d\theta} = 100 \cos (\theta - 15°)$, which is a maximum when $(\theta - 15°) = 0$, or $\theta = 15°$; the value of the slope being then $= 100$. When $\theta = 75°$ the slope is $100 \cos (75° - 15°)$ $= 100 \cos 60° = 100 \times \frac{1}{2} = 50$

(8) $\cos \theta \sin 2\theta + 2 \cos 2\theta \sin \theta = 2 \sin \theta (\cos^2\theta + \cos 2\theta)$
$= 2 \sin \theta (3 \cos^2\theta - 1)$

(9) $amn\theta^{n-1} \tan^{m-1} (\theta^n) \sec^2 (\theta^n)$

(10) $e^x (\sin^2 x + \sin 2x)$

(11) (i) $\dfrac{dy}{dx} = \dfrac{ab}{(x + b)^2}$ (ii) $\dfrac{a}{b} e^{-\frac{x}{b}}$ (iii) $\dfrac{1}{90°} \times \dfrac{ab}{(b^2 + x^2)}$

(12) (i) $\dfrac{dy}{dx} = \sec x \tan x$ (ii) $\dfrac{dy}{dx} = -\dfrac{1}{\sqrt{1 - x^2}}$

(iii) $\dfrac{dy}{dx} = \dfrac{1}{1 + x^2}$ (iv) $\dfrac{dy}{dx} = \dfrac{1}{x\sqrt{x^2 - 1}}$

(v) $\dfrac{dy}{dx} = \dfrac{\sqrt{3 \sec x} \, (3 \sec^2 x - 1)}{2}$

(13) $\dfrac{dy}{d\theta} = 4.6 (2\theta + 3)^{1.3} \cos (2\theta + 3)^{2.3}$

(14) $\dfrac{dy}{d\theta} = 3\theta^2 + 3\cos(\theta + 3) - \ln 3(\cos\theta \times 3^{\sin\theta} + 3^\theta)$

(15) $\theta = \cot\theta$; $\theta = \pm 0.86$; $y = \pm 0.56$; is max. for $+\theta$, min. for $-\theta$

Exercises XV

(1) $x^2 - 6x^2 y - 2y^2$; $\frac{1}{3} - 2x^3 - 4xy$

(2) $2xyz + y^2 z + z^2 y + 2xy^2 z^2$
$2xyz + x^2 z + xz^2 + 2x^2 yz^2$
$2xyz + x^2 y + xy^2 + 2x^2 y^2 z$

(3) $\dfrac{1}{r}\{(x - a) + (y - b) + (z - c)\} = \dfrac{(x + y + z) - (a + b + c)}{r}$; $\dfrac{2}{r}$

(4) $dy = v\,u^{v-1}\,du + u^v \ln u\,dv$

(5) $dy = 3\sin v\,u^2\,du + u^3\cos v\,dv$
$dy = u(\sin x)^{u-1}\cos x\,dx + (\sin x)^u \ln \sin x\,du$
$dy = \dfrac{1}{v}\dfrac{1}{u}\,du - \ln u\,\dfrac{1}{v^2}\,dv$

(7) There is no minimum or maximum.

(8) (*a*) Length 2 feet, width = depth = 1 foot, vol. = 2 cubic feet

    (*b*) Radius = $\dfrac{2}{\pi}$ feet = 7.64 in., length = 2 feet,
vol. = 2.55

(9) All three parts equal; the product is maximum.

(10) Minimum = e^2 for $x = y = 1$

(11) Min. = 2.307 for $x = \frac{1}{2}$, $y = 2$

(12) Angle at apex = 90°; equal sides = length = $\sqrt[3]{2V}$

Exercises XVI

(1) $1\frac{1}{3}$.   (2) 0.6345   (3) 0.2624

(4) $y = \frac{1}{8}x^2 + C$ (5) $y = x^2 + 3x + C$

Exercises XVII

(1) $\dfrac{4\sqrt{a}\,x^{\frac{3}{2}}}{3}+C$

(2) $-\dfrac{1}{x^3}+C$

(3) $\dfrac{x^4}{4a}+C$

(4) $\frac{1}{3}x^3+ax+C$

(5) $-2x^{-\frac{5}{2}}+C$

(6) $x^4+x^3+x^2+x+C$

(7) $\dfrac{ax^2}{4}+\dfrac{bx^3}{9}+\dfrac{cx^4}{16}+C$

(8) $\dfrac{x^2+a}{x+a}=x-a+\dfrac{a^2+a}{x+a}$ by division. Therefore the answer

is $\frac{1}{2}x^2-ax+a(a+1)\ln(x+a)+C$

(9) $\dfrac{x^4}{4}+3x^3+\dfrac{27}{2}\,x^2+27x+C$

(10) $\dfrac{x^3}{3}+\dfrac{2-a}{2}\,x^2-2ax+C$

(11) $a^2\left(2x^{\frac{3}{2}}+\frac{9}{4}x^{\frac{4}{3}}\right)+C$

(12) $-\frac{1}{3}\cos\theta-\frac{1}{6}\theta+C$

(13) $\frac{1}{2}\theta+\dfrac{\sin 2a\theta}{4a}+C$

(14) $\frac{1}{2}\theta-\frac{1}{4}\sin 2\theta+C$

(15) $\frac{1}{2}\theta-\dfrac{\sin 2a\theta}{4a}+C$

(16) $\frac{1}{3}e^{3x}+C$

(17) $\ln|1+x|+C$

(18) $-\ln|1-x|+C$

Exercises XVIII

(1) Area $=120$; mean ordinate $=20$.

(2) Area $=\frac{4}{3}a^{\frac{3}{2}}$

(3) Area $=2$; mean ordinate $=2/\pi=0.637$

(4) Area $=1.57$; mean ordinate $=0.5$

(5) $0.571,\ 0.0476$

(6) Volume $=\frac{1}{3}\pi r^2 h$

(7) 1.25

(8) 79.6

(9) Volume $=4.935$, from 0 to π

(10) $a\ln a,\ \dfrac{a}{a-1}\ln a$

(12) A.M. = 9.5; Q.M. = 10.85

(13) Quadratic mean = $\dfrac{1}{\sqrt{2}} \sqrt{A_1{}^2 + A_3{}^2}$;

arithmetical mean = 0.

The first involves the integral

$$\int (A_1{}^2 \sin^2 x + 2A_1A_3 \sin x \sin 3x + A_3{}^2 \sin^2 3x)dx$$

which may be evaluated by putting $\sin^2 x = \frac{1}{2}(1 - \cos 2x)$, $\sin^2 3x = \frac{1}{2}(1 - \cos 6x)$ and $2 \sin x \sin 3x = \cos 2x - \cos 4x$.

(14) Area is 62.6 square units. Mean ordinate is 10.43.

(16) 436.3 (This solid is pear-shaped.)

Exercises XIX

(1) $\dfrac{x\sqrt{a^2 - x^2}}{2} + \dfrac{a^2}{2} \arcsin \dfrac{x}{a} + C$ (2) $\dfrac{x^2}{2} \left(\ln x - \frac{1}{2}\right) + C$

(3) $\dfrac{x^{a+1}}{a+1} \left(\ln x - \dfrac{1}{a+1}\right) + C$ (4) $\sin e^x + C$

(5) $\sin (\ln x) + C$ (6) $e^x(x^2 - 2x + 2) + C$

(7) $\dfrac{1}{a+1} (\ln x)^{a+1} + C$ (8) $\ln |\ln x| + C$

(9) $2 \ln |x - 1| + 3 \ln |x + 2| + C$

(10) $\frac{1}{2} \ln |x - 1| + \frac{1}{5} \ln |x - 2| + \frac{3}{10} \ln |x + 3| + C$

(11) $\dfrac{b}{2a} \ln \left| \dfrac{x - a}{x + a} \right| + C$ (12) $\ln \left| \dfrac{x^2 - 1}{x^2 + 1} \right| + C$

(13) $\frac{1}{4} \ln \left| \dfrac{1 + x}{1 - x} \right| + \frac{1}{2} \arctan x + C$ (14) $-\dfrac{\sqrt{a^2 - b^2x^2}}{b^2} + C$

Exercises XX

(1) $T = T_0 e^{\mu\theta}$ (2) $s = ut + \frac{1}{2}at^2$

(3) Multiplying out by e^{2t} gives $\dfrac{d}{dt}(ie^{2t}) = e^{2t}\sin 3t$, so that,

$$ie^{2t} = \int e^{2t}\sin 3t\, dt = \tfrac{1}{13}e^{2t}(2\sin 3t - 3\cos 3t) + E$$

Since $i = 0$ when $t = 0$, $E = \tfrac{3}{13}$; hence the solution becomes $i = \tfrac{1}{13}(2\sin 3t - 3\cos 3t + 3e^{-2t})$.

Exercises XXI

(1) $r = 2\sqrt{2}$, $x_1 = -2$, $y_1 = 3$ (2) $r = 2.83$, $x_1 = 0$, $y_1 = 2$

(3) $x = \pm 0.383$, $y = 0.147$ (4) $r = \sqrt{2|m|}$, $x_1 = y_1 = 2\sqrt{m}$

(5) $r = 2a$, $x_1 = 2a$, $y_1 = 0$

(6) When $x = 0$, $r = y_1 = $ infinity, $x_1 = 0$
When $x = +0.9$, $r = 3.36$, $x_1 = -2.21$, $y_1 = +2.01$
When $x = -0.9$, $r = 3.36$, $x_1 = +2.21$, $y_1 = -2.01$

(7) When $x = 0$, $r = 1.41$, $x_1 = 1$, $y_1 = 3$
When $x = 1$, $r = 1.41$, $x_1 = 0$, $y_1 = 3$
Minimum $= 1.75$

(8) For $x = -2$, $r = 112.3$, $x_1 = 109.8$, $y_1 = -17.2$
For $x = 0$, $r = x_1 = y_1 = $ infinity
For $x = 1$, $r = 1.86$, $x_1 = -0.67$, $y_1 = -0.17$

(9) $x = -0.33$, $y = +1.07$

(10) $r = 1$, $x = 2$, $y = 0$ for all points. A circle.

(11) When $x = 0$, $r = 1.86$, $x_1 = 1.67$, $y_1 = 0.17$
When $x = 1.5$, $r = 0.365$, $x_1 = 1.59$, $y_1 = 0.98$
$x = 1$, $y = 1$ for zero curvature.

(12) When $\theta = \dfrac{\pi}{2}$, $r = 1$, $x_1 = \dfrac{\pi}{2}$, $y_1 = 0$.

When $\theta = \dfrac{\pi}{4}$, $r = 2.598$, $x_1 = 2.285$, $y_1 = -1.414$

(14) When $\theta = 0$, $r = 1$, $x_1 = 0$, $y_1 = 0$

When $\theta = \dfrac{\pi}{4}$, $r = 2.598$, $x_1 = -0.715$, $y_1 = -1.414$

When $\theta = \dfrac{\pi}{2}$, $r = x_1 = y_1 = \text{infinity}$

(15) $r = \dfrac{(a^4 y^2 + b^4 x^2)^{\frac{3}{2}}}{a^4 b^4}$, when $x = 0$, $y = \pm b$, $r = \dfrac{a^2}{b}$,

$x_1 = 0$, $y_1 = \pm \dfrac{b^2 - a^2}{b}$; when $y = 0$, $x = \pm a$, $r = \dfrac{b^2}{a}$,

$x_1 = \pm \dfrac{a^2 - b^2}{a}$, $y_1 = 0$

(16) $r = 4a \, |\sin \tfrac{1}{2}\theta|$

Exercises XXII

(1) $s = 9.487$ (2) $s = (1 + a^2)^{\frac{3}{2}}$ (3) $s = 1.22$

(4) $s = \displaystyle\int_0^2 \sqrt{1 + 4x^2} \, dx$

$= \left[\dfrac{x}{2} \sqrt{1 + 4x^2} + \tfrac{1}{4} \ln \left(2x + \sqrt{1 + 4x^2} \right) \right]_0^2 = 4.65$

(5) $s = \dfrac{0.57}{m}$ (6) $s = a(\theta_2 - \theta_1)$ (7) $s = \sqrt{r^2 - a^2}$

(8) $s = \displaystyle\int_0^a \sqrt{1 + \dfrac{a}{x}} \, dx$ and $s = a\sqrt{2} + a \ln \left(1 + \sqrt{2}\right) = 2.30a$

(9) $s = \dfrac{x-1}{2} \sqrt{(x-1)^2 + 1} + \frac{1}{2} \ln \left\{ (x-1) + \sqrt{(x-1)^2 + 1} \right\}$

and $s = 6.80$

(10) $s = \displaystyle\int_1^e \dfrac{\sqrt{1+y^2}}{y} \, dy.$ Put $u^2 = 1 + y^2$; this leads to

$$s = \sqrt{1+y^2} + \ln \dfrac{y}{1 + \sqrt{1+y^2}} \text{ and } s = 2.00$$

(11) $s = 4a \displaystyle\int_0^\pi \sin \dfrac{\theta}{2} \, d\theta$ and $s = 8a$

(12) $s = \displaystyle\int_0^{\frac{1}{4}\pi} \sec x \, dx.$ Put $u = \sin x$; this leads to

$s = \ln \left(1 + \sqrt{2}\right) = 0.8814$

(13) $s = \dfrac{8a}{27} \left\{ \left(1 + \dfrac{9x}{4a}\right)^{\frac{3}{2}} - 1 \right\}$

(14) $s = \displaystyle\int_1^2 \sqrt{1 + 18x} \, dx.$ Let $1 + 18x = z$, express s in terms

of z and integrate between the values of z corresponding

to $x = 1$ and $x = 2$. $s = 5.27$

(15) $s = \dfrac{3a}{2}$ (16) $4a$

All earnest students are exhorted to manufacture more examples for themselves at every stage, so as to test their powers. When integrating he can always test their answer by differentiating it, to see whether they gets back the expression from which they started.

APPENDIX

SOME RECREATIONAL PROBLEMS RELATED TO CALCULUS

———✸———

THE OLD OAKEN CALCULUS PROBLEM
How dear to my heart are cylindrical wedges,
when fond recollection presents them once more,
and boxes from tin by upturning the edges,
and ships landing passengers where on the shore.
The ladder that slid in its slanting projection,
the beam in the corridor rounding the ell,
but rarest of all in that antique collection
the leaky old bucket that hung in the well—
the leaky old bucket, the squeaky old bucket,
the leaky old bucket that hung in the well.

Katherine O'Brien, in
The *American Mathematical Monthly*
(vol. 73, 1966, p. 881).

Problems solvable only by calculus are extremely rare in popular puzzle books. One exception is a puzzle that usually involves an animal A that is a certain distance from a man or another animal B. Assume the animal is a cat directly north of a dog. Both run at a constant rate, but B goes faster than A. If both animals run due north, the dog will of course eventually catch the cat. In this form it is equivalent to Zeno's famous paradox of Achilles and the tortoise. It is the simplest example of what are called "pursuit paths." Finding the distance traveled by pursuer and pursued, in this linear form, is an easy task involving only arithmetic.

Pursuit paths on the plane become more interesting. Assume that the cat travels in a straight line due east, and that the dog always runs directly toward the cat. Both go at a constant rate. If the dog is faster than the cat it can always catch the cat. Given its initial distance south of the cat, and the ratio of their speeds, how far does the cat go before it is caught? How far does the dog go, along its curved path, before it catches the cat?

Versions of this two-dimensional pursuit problem are not quite so easy to solve. Such puzzles were popular in the eighteenth century when they usually involved one ship pursuing another. Later versions took the form of running animals and persons. I will cite two examples from classic twentieth-century puzzle collections.

Henry Dudeney, in *Puzzles and Curious Problems* (1931, Problem 210) describes the situation this way. Pat is 100 yards south of a pig. The pig runs due west. Pat goes twice as fast as the pig. If he always runs directly toward the pig, how far does each go before the pig is caught?

Sam Loyd, in his *Cyclopedia of Puzzles* (1914, p. 217), also makes the pursued a pig, but now it is being chased by Tom the Piper's Son. (He stole a pig, remember, in the old Mother Goose rhyme). Tom is 250 yards south of the pig and the pig runs due east. Both go at uniform rates with Tom running $\frac{4}{3}$ as fast as the pig. Tom always runs directly toward the pig. Again, how far does each travel before the pig is caught?

Such problems can be solved the hard way by integrating. Dudeney answers his puzzle with no explanation of how to solve it. He adds: "The curve of Pat's line is one of those curves the length of which may be exactly measured. But we have not space to go into the method."

Like so many calculus problems, it turns out that pursuit problems of this type often can be handled by simple formulas, although calculus is needed to prove them. One such method is given by L.A. Graham in his *Ingenious Mathematical Problems and Methods* (1959, Problem 74). His version of the puzzle involves a dog chasing a cat. The dog, 60 yards south of the cat, runs $\frac{5}{4}$ as fast as the cat who runs due east. Graham first solves the problem by integrating, then provides the following surprising formula. The distance traveled by the cat equals the initial distance between dog and cat multiplied by the fraction that expresses the ratio of their speeds, divided by a number one less than the square of the ratio.

Using the parameters of Graham's problem, the distance traveled by the cat is:

$$\frac{60 \times 5}{4} \div \left(\frac{5^2}{4^2} - 1\right)$$

which works out to 133 and $\frac{1}{3}$ yards. Because the dog goes $\frac{5}{4}$ as fast as the cat, it travels $\frac{5}{4}$ times 133 and $\frac{1}{3}$, or 166 and $\frac{2}{3}$ yards.

Applying the formula to Dudeney's version shows that the pig runs 66 and $\frac{2}{3}$ yards, and Pat, who goes twice as fast, runs 133 and $\frac{1}{3}$ yards. In Loyd's version the pig goes 428 and $\frac{4}{7}$ yards. Tom travels $\frac{4}{3}$ times that distance, or 571 and $\frac{3}{7}$ yards.

Another well known pursuit path problem, covered in many books on recreational mathematics, involves n bugs at the corners of a regular n-sided polygon of unit side. The bugs simultaneously start to crawl directly toward their nearest neighbor on, say, the left. (It doesn't matter whether they go clockwise or counterclockwise). All bugs move at the same constant rate. It is intuitively obvious that they will travel spiral paths that come together at the polygon's center. When they all meet, how far has each bug traveled?

Although the problem can be posed with any regular polygon, it is usually based on a square. At any given moment the four bugs will be at the corners of a square that steadily diminishes and rotates until the four bugs meet at the center. Their paths are logarithmic spirals. Although the length of each path can be determined by calculus, the problem can be solved quickly without calculus if you have the following insight.

At all times the path of each bug is perpendicular to the bug it is pursuing. It follows that there is no component of a pursued bug's motion that carries it toward or away from its pursuer. Consequently, the pursuer will reach the pursued in the same time it would take if the pursued bug remained stationary and its pursuer moved directly toward it. Each spiral path will therefore have the same length as the side of the unit square.

A geometrical way of measuring the distance traveled by each bug starting at the corner of a regular polygon is shown in Figure 11 where it is applied to a regular hexagon. Draw line AO from a corner to the polygon's center, then extend OX at right angles to AO until it meets either side AB or its extension as shown. AX, which is r times the secant of angle theta, is then the length of the path traveled by each bug. In the case of the square, AX is the square's side. In the case of a triangle, OX meets AB at a spot two-thirds of the way from A, indicating that the bug travels two-thirds of the length of the triangle's side. For the hexagon, theta is $60°$, and the bug travels twice the distance of a side.

Minimum path problems that require differentiating occasionally turn up in books on puzzles as well as in calculus textbooks. They usually take the form of a swimmer who is in a lake at distance x from a straight shoreline. He wants to get to a certain point p along the coast. Given a steady speed at which he swims, and his steady speed while walking on land, what spot along the coast should he swim to so as to minimize the total time it takes him to swim to shore and then walk to point p along the coast?

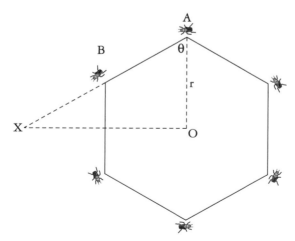

FIG. 11. How to calculate a bug's pursuit path on a regular hexagon.

In his *Cyclopedia of Puzzles* (page 165), Sam Loyd offers a version of such a problem. His text and its accompanying illustration are shown in Figure 12. Loyd's constants make the problem tedious to calculate, but the solution by equating a derivative to zero is straightforward.

FIG. 12. Sam Loyd's steeplechase puzzle from his *Cyclopedia of Puzzles.* Here is a little cross-country steeplechase problem which developed during the recent meeting, which will interest turfites as well as puzzlists. It appears that toward the end of a well-contested course, when there was but a mile and three quarters yet to run, the leaders were so closely bunched together that victory turned upon the selection of the best or shortest road. The sketch shows the judges' stand to be at the opposite end of a rectangular field, bounded by a road of a mile long on one side by three-quarters of a mile on the other.

By the road, therefore, it would be a mile and three-quarters, which all of the horses could finish in three minutes. They are at liberty, however, to cut across lots at any point they wish, but over the rough ground they could not go so fast. So while they would lessen the distance, they would lose twenty-five per cent. in speed. By going directly across on the bias, or along the line of the hypotenuse as the mathematicians would term it, the distance would be a mile and a quarter exactly. What time can the winner make by selecting the most judicious route?

Figure 13 diagrams the relevant rectangle. Let x be the
distance from the jump point to
the rectangle's corner B, and
$(1 - x)$ be the distance from
where the horses start to where
they should jump the wall. The
path the horses take on rough
ground is the hypotenuse of a
right triangle with a length
equal to the square root of $(x^2 +$
$.75^2)$. The horses run on the

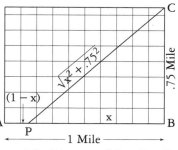

FIG. 13. Diagram of Sam Loyd's
steeplechase puzzle.

smoother ground at a speed of
1.75/3 = .58333+ miles per minute. Their speed on rougher
ground is a fourth less, or .4375 miles per minute.

To solve the problem we must first write an expression for the
total time as a function of x. The time it takes the horses to go
on rough ground from P to C is:

$$\frac{\sqrt{x^2 + .75^2}}{.4375}$$

The time it takes them to cover distance $1 - x$, on the
smoother ground, to the point where they jump the wall is:

$$\frac{1 - x}{.5833}$$

The total time is the sum of these two expressions:

$$t = \frac{\sqrt{x^2 + .75^2}}{.4375} + \frac{1 - x}{.5833}$$

We next differentiate by the "sum rule" (see Chapter VI):

$$\frac{dt}{dx} = \frac{x}{.4375\sqrt{x^2 + .75^2}} - \frac{1}{.5833}$$

We equate the derivative to zero and solve for x. This gives x
a value of .8504+ miles. Subtracting this from 1 gives the dis-

tance on smooth ground, from the start to the jump point, as about .15 miles. The times are easily obtained by substituting .8504 for x in the two time expressions. The total time is 2.85 minutes or about 2 minutes and 51 seconds.

Because the ratio of the two speeds is $\frac{2.25}{3}$ or $\frac{3}{4}$, we can greatly simplify the calculation by substituting 3 and 4 for the respective velocities:

$$\frac{\sqrt{x^2 + .75^2}}{3} + \frac{1-x}{4}$$

The derivative of this function becomes:

$$\frac{x}{3\sqrt{x^2 + .75^2}} - \frac{1}{4}$$

Equate to zero:

$$\frac{x}{3\sqrt{x^2 + .75^2}} - \frac{1}{4} = 0$$

$$\frac{x}{3\sqrt{x^2 + .75^2}} = \frac{1}{4}$$

Square both sides:

$$\frac{x^2}{9(x^2 + .75^2)} = \frac{1}{16}$$

$$16x^2 = 9(x^2 + .75^2) = 9x^2 + 5.0625$$

$$16x^2 - 9x^2 = 5.0625$$

$$7x^2 = 5.0625$$

$$x^2 = 5.0625/7 = .72321428 \ldots$$

$$x = .85042 \ldots$$

Occasionally what seems to be a very difficult path problem, involving the sum of an infinite converging series, can be solved in a flash if you have the right insight. A classic instance is the brainteaser about two locomotives that face each other on the same track, 100 miles apart. Each locomotive travels at a speed of 50 miles per hour. On the front of one engine is a fly that flies back and forth between the two engines on a zigzag path that ends when the locomotives collide. If the fly's speed is 80 miles per hour, how far does it travel before the trains crash?

It is not easy to sum the infinite series of zigzags, but this is not necessary. The trains collide in one hour, so if the fly's speed is 80 miles per hour, it will have gone 80 miles.

There is an anecdote about how the great mathematician John von Neumann solved this problem. He is said to have thought for a moment before giving the right answer. "Correct," said the proposer, "but most people think you have to sum an infinite series." Von Neumann looked surprised and said that was how he solved it!

A joke version of this problem gives the fly a normal speed of 40 miles per hour instead of 80. How far does it go? The answer is *not* 40 miles!

In Chapter 8 of my *Wheels, Life, and Other Mathematical Amusements* I published a peculiar path problem invented by A.K. Austin, a British mathematician. Here is how he phrased it:

> "A boy, a girl and a dog are at the same spot on a straight road. The boy and the girl walk forward—the boy at four miles per hour, the girl at three miles per hour. As they proceed, the dog trots back and forth between them at 10 miles per hour. Assume that each reversal of its direction is instantaneous. An hour later, where is the dog and which way is it facing?"
>
> Answer: "The dog can be at any point between the boy and the girl, facing either way. Proof: At the end of one hour, place the dog anywhere between the boy and the girl, facing in either direction. Time-reverse all motions and the three will return at the same instant to the starting point."

The problem generated considerable controversy centering around the question of whether the boy, girl, and dog could ever get started. Philosopher of science Wesley Salmon wrote a *Scientific American* article about it which you'll find reprinted in the book previously cited. The problem is a strange instance of an event that can be precisely defined in forward time, but becomes ambiguous when time is reversed.

Another example of such a paradox involves a spiral curve on a sphere. Imagine a point starting at the earth's equator and moving northeast at constant speed. It traces a spiral path called a loxodrome. The point will circle the north pole an infinite number of times before it finally, after a finite time, strangles the pole as the limit of its path. The point starts from a precise spot on the equator. But if the event is time reversed, the point can end at *any* spot on the equator. Are the time reversed versions of the loxodrome and Austin's dog self-contradictory because they require starting an infinite series at its limit? Can the two situations be resolved by applying nonstandard analysis—a form of calculus mentioned in my chapter on limits? For more details, see Professor Salmon's article.

Limits can lead to a variety of mind-twisting fallacies and paradoxes. Consider the doubling series $x = 1 + 2 + 4 + 8 + \ldots$. Apply to it the trick I explained in my chapter on limits. Multiply each side by 2:

$$2x = 2 + 4 + 8 + 16 + \ldots.$$

The right side clearly is the original series minus 1, therefore $2x = x - 1$. We seem to have proved that the series has a final sum of -1.

A less obvious fallacy occurs when we consider the following two series, each with terms that get progressively smaller so it seems as if each converges:

$$x = \tfrac{1}{1} + \tfrac{1}{3} + \tfrac{1}{5} + \tfrac{1}{7} + \ldots.$$
$$y = \tfrac{1}{2} + \tfrac{1}{4} + \tfrac{1}{6} + \tfrac{1}{8} + \ldots.$$

The first series is the harmonic series with all even denominator fractions dropped. The second is the harmonic series with all odd denominator fractions omitted.

Multiply both sides of the second series by 2:

$$2y = \tfrac{1}{1} + \tfrac{1}{2} + \tfrac{1}{3} + \tfrac{1}{4} + \ldots.$$

This is the harmonic series. Clearly it is the sum of x and y. If $x + y = 2y$, then $x = y$, and we seem to have shown that $x - y = 0$.

We now group the terms of the harmonic series so that each even denominator fraction is subtracted from an odd denominator one:

$$x - y = (1 - \tfrac{1}{2}) + (\tfrac{1}{3} - \tfrac{1}{4}) + (\tfrac{1}{5} - \tfrac{1}{6}) + \ldots$$

Each term inside parentheses is greater than zero. In other words $x - y$ is greater than zero, so apparently we have shown that 0 is greater than 0.

One encounters similar fallacies when terms in a nonconverging series are grouped in various ways.

An even wilder fallacy, which bewildered many mathematicians in the early days of calculus, involves the oscillating series:

$$1 - 1 + 1 - 1 + 1 - 1 + \ldots$$

If terms are grouped like this:

$x = (1 - 1) + (1 - 1) + (1 - 1) + \ldots$, then the series becomes $0 + 0 + 0 + \ldots$ and $x = 0$.

But if we group like so:

$$x = 1 - (1 - 1) - (1 - 1) - (1 - 1) - \ldots$$
$$x = 1 - 0 - 0 - 0 - \ldots$$
$$x = 1$$

Leonhard Euler, by the way, argued that the sum was $\tfrac{1}{2}$.

Mathematicians distinguish between infinite series that are *absolutely* convergent and those that are *conditionally* convergent. If grouping or rearranging terms has no effect on a series' limit, the series is absolutely convergent. This is always the case if in a series with a mix of plus and minus signs the plus terms taken separately converge, and so do the terms with minus signs taken separately. For example: $1 - \tfrac{1}{2} + \tfrac{1}{4} - \tfrac{1}{8} + \ldots$ is absolutely convergent, with a limit of $\tfrac{2}{3}$.

The series $1 - \tfrac{1}{2} + \tfrac{1}{3} - \tfrac{1}{4} + \ldots$ is conditionally convergent because its plus and minus terms, taken separately, do not converge. By rearranging its terms it can have any limit you please. If the terms are not rearranged, their partial sums hop back and forth to finally converge on $.693147 \ldots$, the natural logarithm of 2.

An ancient geometrical paradox involving a series of line segments that appear to reach a limit, but do not, is the fallacious

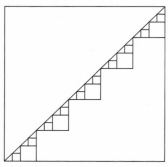

FIG. 14. "Proof" that a square's diagonal equals the sum of its two sides.

"proof" that a square's diagonal is equal to the sum of its two sides. Figure 14 shows a series of stairsteps along a square's diagonal. Because each new series of steps is smaller than the previous series, it looks as if the stairsteps eventually get so tiny that they become the diagonal. Because each stairstep has a length equal to the sum of two sides of the square, have we not shown that the diagonal is the sum of two sides?

Of course we haven't. It is true that the stairsteps, as they get smaller, converge on the diagonal as a limit, but this is a case where they never actually reach the limit. No matter how far we carry the procedure, the total length of the steps remains twice the square's side.

In a similar way we can "prove" that half the circumference of a circle equals the circle's diameter. The "proof" is based on the yin-yang symbol of the Orient, as shown in Figure 15. As the semicircles get smaller and smaller they appear to become the circle's diameter. However, as in the previous fallacy, they never actually reach this limit. The construction along the circle's diameter, and the construction along the square's diagonal, are fractals. Magnify any portion of either "curve" and it always looks the same.

A beautiful example of a fractal of infinite length even though it surrounds a finite area is the snowflake curve. It is produced by attaching a small equilateral triangle to the central third of each side of an equilateral triangle, then continuing to add smaller and smaller triangles to the new sides. Imagine this process carried to infinity. When the snowflake reaches its limit, its perimeter

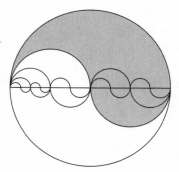

FIG. 15. A yin-yang "proof" that half a circle's circumference equals its diameter.

becomes infinitely long. At the limit it has no tangent at any point, therefore it is a curve without a derivative. Figure 16 shows what the curve looks like after a small number of steps.

When I wrote about the snowflake in my *Scientific American* column I said it had a solid analog with very similar properties. Imagine a regular tetrahedron with a smaller regular tetrahedron attached to the central fourth of each face. Assume this procedure is repeated, with smaller and smaller tetrahedra, to infinity. The resulting surface, I wrote, is a fractal surface, crinkly like the snowflake, infinite in area even though it surrounds a finite volume of space.

My intuition was wrong. William Gosper, a computer scientist famed for his discovery of the glider gun in John Conway's fabled game of Life, was suspicious of my remarks. He wrote a computer program to trace the progress of the solid snowflake. He found that at the limit the polyhedron converged on the surface of a cube! Although the cube is crosscrossed with an infinity of lines, at the limit the lines have no thickness, so the cube's surface is perfectly smooth.

Another striking example of a solid with a finite volume but an infinite surface area is provided by the sequence of attached cubes shown in Figure 17. The top cube has a side of 1. The sides di-

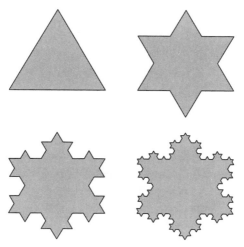

FIG. 16. How the snowflake curve grows.

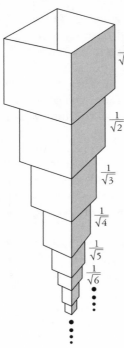

minish in the series: $1 + \dfrac{1}{\sqrt{2}} + \dfrac{1}{\sqrt{3}} + \dfrac{1}{\sqrt{4}} + \ldots + \dfrac{1}{\sqrt{n}} + \ldots$

The series diverges, which means that the polyhedron grows to an infinite length. The sum of the areas of the faces also diverges. Consider only the faces shown shaded. Their areas are in the series: $1 + \frac{1}{2} + \frac{1}{3} + \frac{1}{4} + \ldots + \frac{1}{n} \ldots$

Do you recognize it? It is the harmonic series which we know diverges. It would require an infinite supply of paint just to paint one side of each cube! On the other hand, the volumes of the cubes fall into the series:

$$1 + \dfrac{1}{2\sqrt{2}} + \dfrac{1}{3\sqrt{3}} + \dfrac{1}{4\sqrt{4}} + \ldots + \dfrac{1}{n\sqrt{n}} + \ldots$$

FIG. 17. A solid of finite volume but infinite surface area.

This series converges. The total volume of the cubes is a finite number, but the sum of their surface areas is infinite!

EXTREMUM PROBLEMS

Examples of extrema (maximum and minimum values) abound in physics. Soap films form minimal surface areas. Refracted light minimizes the time it takes to go from A to B. In general relativity bodies moving freely in space-time follow geodesics (shortest paths), and planets and stars seek to minimize their surface area, though mountains and rotational bulges go the other way. Examples are endless. The fact that at extreme points in functions their derivatives equal zero is a striking tribute to the simplicity which nature seems to favor in its fundamental laws.

A maximum value of one variable in a function often minimizes another variable, and vice versa. A circle's circumference, for instance, is a minimum length that surrounds a given area.

Conversely, the circle is the closed curve of given length that maximizes the area it surrounds. A miniskirt is designed both to minimize a skirt's length and maximize the amount of exposed legs. A maxiskirt maximizes its length and minimizes the exposed legs.

In Chapter XI Thompson asks how a number n should be divided into two parts so that the product of the parts is maximum. He shows how easy it is to find the answer by equating a derivative to zero. The number must be divided in half, giving a product equal to $n^2/4$.

A father tosses a handful of coins on a table and says to his son: "Separate those coins into two parts, and your weekly allowance will be the product of the amount in each part. The son maximizes his allowance by dividing the coins so that the two amounts are as nearly equal as possible.

The problem of dividing n into two parts to get a maximum product is equivalent to the following geometrical problem. Given a rectangle's perimeter, what sides will maximize its area? The answer of course is a square. For example, suppose the perimeter is 14. The sum of the two adjacent sides is 7. To maximize the rectangle's area each side must be half of 7 or 3.5, forming a square of area 12.25.

Paul Halmos, in his *Problems for Mathematicians Young and Old* (1991), asks in Problem 51 for the shortest curve that bisects the area of an equilateral triangle. You might guess it to be a straight line parallel to a side, a line of length .707. . . . The correct answer is a circular arc of length .673. . . . You can prove this with calculus, but there is an easier way. Consider the regular hexagon shown in Figure 18. The figure of smallest perimeter that bisects the hexagon's area is the circle shown. Its arc within each of the six equilateral triangles must therefore bisect each triangle. The hexagon's area, assuming its side is 1, is $3\sqrt{3}/2$. The circle of half that area is $3\sqrt{3}/4 = 1.299$. . . . It is not hard

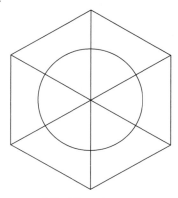

FIG. 18. The shortest curve bisecting an equilateral triangle.

now to find its radius to be .643 . . . , and the length of the arc that bisects each triangle to be

$$\frac{\pi}{3} \sqrt{\frac{3\sqrt{3}}{4\pi}} = .673. \ldots$$

The proof of course assumes the fact that a circle is the closed curve of given length that maximizes its interior area.

Thompson extends his proof that a number divided into two equal parts gives a maximum product to a proof that if a number is broken into *n* parts, the maximum product of the parts results when all *n* parts are equal. For example, if the number is 15, the 3-part partition with a maximum product is $5 \times 5 \times 5 = 125$.

This provides a neat proof that given the perimeter of a triangle, the largest area results if the triangle is equilateral. The proof rests on an elegant formula for determining the area of any triangle when given its three sides. It is called Hero's (or Heron's) formula after Hero (Heron) of Alexandria, an ancient Greek mathematician. (The formula has several algebraic proofs, but they are complicated.) Let *a,b,c* stand for a triangle's sides, and *s* for its semiperimeter (half the perimeter). Hero's famous formula is:

$$Area = \sqrt{s(s - a)(s - b)(s - c)}$$

It is easy to see that the area is maximized when the three values inside the parentheses are equal. This is true only if $a = b = c$, making the triangle equilateral. If each side is 1, the triangle's area will be $\sqrt{3}/4$.

Suppose we are given the perimeter of a triangle and its base. What sides will maximize its area? Calculus will tell you. Here's another way. Imagine two pins separated by the distance of the triangle's base. A string with a length equal to the perimeter circles the pins as shown in Figure 19. A pencil point at *c* keeps the string taut. As you move the pencil point from side to side, clearly the highest altitude of the triangle formed by the string is obtained when the sides are equal, making the triangle isosceles. (Continuing to move the pencil around the pins will trace an ellipse.)

Assume that the sides of an isosceles triangle remain constant, but the angle they make between them is allowed to vary. This of

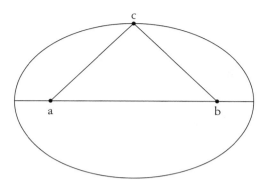

FIG. 19. A string proof of a maximum problem.

course also varies the length of the triangle's third side. What angle maximizes the triangle's area? Knowing that for a given perimeter the equilateral triangle has the largest area, one is tempted to guess that the angle will be 60 degrees. Surprisingly, this is not the case. The maximizing angle is 90 degrees, making the triangle an isosceles right triangle.

Calculus will prove this, but here is a simpler proof. Consider the isosceles right triangle shown on its side in Figure 20. The equal sides *a* and *b* are fixed. By moving the top vertex (c) left or right, keeping *b* the same length, we can vary the angle theta. Note that regardless of which direction the vertex moves, the altitude of the triangle is shortened. Because a triangle's area is half

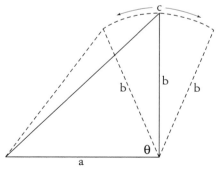

FIG. 20. A simple proof of an isosceles triangle problem.

the product of its base and altitude, the largest area results when the altitude is maximum. This occurs when the angle theta is neither acute nor obtuse, but a right angle.

Another simple proof that the angle is 90° is provided by imagining that the triangle's variable side is a mirror. As Figure 21 shows, the reflection produces a rhombus. As we have learned (and Thompson proves by calculus in Chapter XI), the rectangle of given perimeter which has the largest area is a square. Figure 21B shows that when the reflection forms a square, the area of the isosceles triangle is maximized by having a right angle between its two equal sides.

A farmer wishes to build a fence of three sides to enclose a rectangular plot of land with a fixed wall as its fourth side. What three lengths of the fence will maximize the plot's area? This problem is a favorite of calculus textbooks, but knowing that a square maximizes a rectangle's area, given the perimeter, solves the problem quickly. Imagine the wall to be a mirror. The plot of land plus its mirror image is a larger rectangle. The largest area of this larger rectangle results if it is a square. The plot of land is half that square. Its longer side will be twice the length of each of the other two sides.

A related problem can be solved using the correct assumption that a regular polygon of n sides encloses the largest area for its perimeter. If the farmer's fence has n sides, the mirror trick shows that the plot's area is maximized by making the sides half of a regular polygon of $2n$ sides. As the sides of a regular polygon increase, they approach the circle as their limit. (This was how

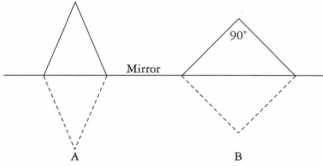

F I G . 2 1 . A mirror reflection proof.

Archimedes found a good approximation for the value of π.) So, if the farmer wishes to surround his plot with a *curved* fence on the side of the wall, the mirror trick shows that he maximizes the plot's area by making the fence the arc of a semicircle.

Sometimes more than one mirror reflection will solve a problem easier than by using calculus. Suppose you have a screen made of two identical halves that are hinged together so you can vary the angle between them. You wish to place them at the corner of a room so as to maximize the enclosed area. Figure 22 shows how two mirrors solve the problem. A polygon of eight equal sides has the largest area when it is regular. The screens must be placed so they form a quadrant of the regular octagon which has angles of 135°.

A maximum problem given in most calculus textbooks (I'm surprised Thompson does not include it) concerns a square to be cut and folded to make a square box without a lid. This is done by cutting out little squares at the four corners of the large square, then folding up the rectangular sides. The question is: What sizes should the removed squares be to produce an open box of maximum volume?

For example, suppose the square is 12 inches on the side. Let x be the side of each corner square to be removed. The box's square base will have a side of $12 - 2x$, or $2(6 - x)$. The area of the square base $= [2(6 - x)]^2 = 4(6 - x)^2$. The volume of the box is x times the area of the base, or $4x(6 - x)^2$.

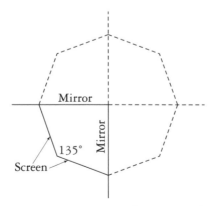

FIG. 22. A mirror reflection proof.

The derivative simplifies to $12(6 - x)(2 - x)$. When this is equated to zero, we find that x has a value of either 6 or 2. It can't be 6, because cutting out six-inch squares would leave nothing to fold up, therefore $x = 2$. The box of maximum volume has a square base of 8 inches on the side, a depth of 2 inches, and a volume of $2 \times 64 = 128$ cubic inches.

Note that the area of the base is 64 inches, and the total area of the sides is also 64 inches. This equality holds regardless of the size of the original square.

S. King, in "Maximizing a Polygonal Box" (*The Mathematical Gazette,* Vol. 81, March 1997, pp. 96-99) shows that this is true for any convex polygon whose corners are cut and the rectangular sides folded to make an open box. If the polygon is a triangle or a regular polygon the volume is maximized when the ratio of the combined area of the sides to the area of the discarded corners is 4 to 1.

The task of determining the largest square that will go inside a cube is involved in a famous problem known as Prince Rupert's problem after a nephew of England's King Charles. (The Prince lived from 1619 to 1682.) Rupert asked: Is it possible to cut a hole through a cube large enough for a slightly larger cube to be passed through the tunnel?

If you hold a cube so a corner points directly toward you you will see the regular hexagon shown in Figure 23, left. It is possible to inscribe within this hexagon a square that has a side a trifle larger than the cube's face. Figure 24, right, shows the square from a different perspective. Note that two of the

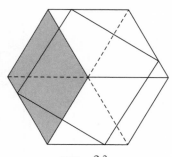

FIG. 23.

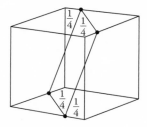

FIG. 24. Two views of Prince Rupert's hole through cube.

square's sides are visible on outside faces of the cube, whereas the other two sides are inside the cube and invisible if the cube is opaque.

The square is the largest possible square that fits inside a cube. Each corner, as indicated, is one-fourth the distance along an edge from a corner. If the cube's edge is 1, the side of the inscribed square is $3/\sqrt{8} = 1.060660.\ \ldots$ This is three-fourth's of the cube's face diagonal, a tiny bit longer than the cube's edge. This allows a tunnel to be cut through a solid cube—a tunnel with the inscribed square as its cross-section. Through the tunnel can be passed a cube whose edge is a trifle less than $1.060660.\ \ldots$ The square's area is exactly $9/8 = 1.125$.

I do not know how to prove by differentiating that this square is the largest that fits in a cube, and would enjoy hearing from any reader who can tell me. The problem generalizes to cubes of higher dimension. It is not easy to determine the largest cube that will fit within a hypercube of four dimensions. For higher dimensions the general problem remains unsolved. See Problem B4 in *Unsolved Problems in Geometry* (1991) by Hallard Croft, Kenneth Falconer, and Richard Guy.

The largest rectangle, by the way, that fits inside a cube is the cross-section that goes through the diagonals of two opposite faces. On the unit cube it is a rectangle of sides $1 \times \sqrt{2}$, with an area of $\sqrt{2}$.

CYLINDERS

Problems about right circular cylinders abound in calculus textbooks. Here is a delightful little-known puzzle that I found in *Problems for Puzzlebusters* (1992, pp 25-26), by David L. Book.

A right circular cylindrical can is open at the top. The can's diameter is 4 inches, its height is 6 inches. A sphere is dropped into the can. The task is to find the size of the ball that maximizes the amount of liquid that must be poured into the can to cover the ball exactly.

You might think that the largest possible ball, four inches in diameter, would do the trick. With more reflection you realize that a smaller ball might require more liquid because there would be more space around it to fill.

Let x be the ball's diameter. The ball, plus the liquid that just covers it, forms a right circular cylinder with a height equal to the ball's diameter. The height of this cylinder (ball plus liquid) is also x. The cylinder's volume is $4\pi x$. The ball's volume is

$$4\pi\left(\frac{x}{2}\right)^3 /3 = \pi x^3/6.$$

The amount of liquid required to cover the ball is the difference between the ball's volume and the cylinder's volume, so we can write:

$$v = 4\pi x - \frac{\pi x^3}{6}$$

$$= \frac{24\pi x}{6} - \frac{\pi x^3}{6}$$

$$= \frac{\pi}{6}\,(24x - x^3)$$

The above expression gives the amount of liquid needed to cover the ball as a function of x, the ball's diameter. To find the largest size of the ball, $\pi/6$ is a constant so it can be dropped. Only $24x - x^3$ need be differentiated and equated to zero. The derivative is $24 - 3x^2$. Equating to zero and solving for x gives the ball's diameter as $\sqrt{8} = 2.828+$.

In Chapter V Thompson considers the rate at which a right circular cylinder's volume varies as a function of its radius. Susan Jane Colley, writing on "Calculus in the Brewery" in *The College Mathematics Journal* (May 1994) points out how manufacturers of products that come in cylindrical cans make use of this function to deceive consumers. She noticed that beer cans which hold exactly the same amount often have different heights but seem to have the same radius. One expects the taller can to hold more beer when actually it does not.

Colley shows by differentiating the formula for the volume of a cylinder (the area of base times height) that if a cylindrical beer can has a very slight decrease in radius, a decrease imperceptible to the eye, it grows taller by ten times the decrease in radius while keeping its volume the same. The taller cans are optical illusions. They seem to hold more beer than the shorter cans, but actually do not. "Smart people, those marketing sales types," Cooley concludes.

Another delightful cylinder problem that deserves to be in calculus textbooks is solved by Aparna W. Higgins in "What is the Lowest Position of the Center of Mass of a Soda Can?", in *Primus* (Vol. 7, March 1997, pp. 35-42). A full can's center of mass is near the can's geometrical center, and the same is true of an empty can. As the drink is consumed, the center of mass steadily lowers. When the can is empty, it is back up to the geometric center. Obviously it cannot go to the bottom, then jump suddenly up to the center, so there must be a point at which the center of mass reaches its lowest point before it starts to rise.

The problem is to find that minimum. Three cylinders are involved: the can, the liquid, and the air above the liquid, assuming, of course, the can is vertical.

Dr. Higgins, a mathematician at the University of Dayton in Ohio, shows how to solve the problem neatly by differentiating and equating to zero. Surprise! The can's center of mass reaches its lowest point when it exactly reaches the level of the liquid!

In Chapter 16 of my *Wheels, Life, and Other Mathematical Amusements* I give this problem along with a reader's clever way of solving it without calculus. However, as Dr. Higgins points out, using differential calculus to find the answer is an excellent exercise for calculus students. I might add that unlike so many textbook problems it is directly related to a student's experience.

Problems of integration that routinely turn up in calculus textbooks can often be solved by simpler methods. A classic example involves two circular cylinders, each with a unit radius, which intersect at right angles as shown in Figure 25, top. What is the volume of the shaded portion that is common to both cylinders?

To answer this question you need know only that the area of a circle is πr^2, and the volume of a sphere is $(4\pi r^3)/3$.

Imagine a sphere of unit radius inside the volume common to the two cylinders and having as its center the point where the axes of the cylinders intersect. Suppose that the cylinders and sphere are sliced in half by a plane through the sphere's center and both axes of the cylinders (See Figure 26, bottom left). The cross section of the volume common to the cylinders will be a

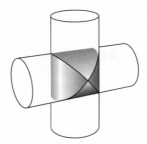

FIG. 25. Archimedes Problem of the crossed cylinders

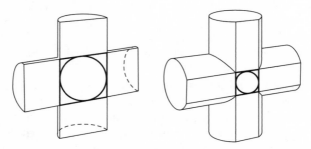

FIG. 26. Crossections of the cylinders.

square. The cross section of the sphere will be a circle inscribed in a square.

Now suppose that the cylinders and sphere are sliced by a plane parallel to the previous one but that shaves off only a small portion of each cylinder (Figure 26, bottom right). This will produce parallel tracks on each cylinder, which intersect as before to form a square cross section of the volume common to both cylinders. Also as before, the cross section of the sphere will be a circle inside the square. It is not hard to see (with a little imagination and pencil doodling) that any plane section through the cylinders, parallel to the cylinders' axes, will always have the same result: a square cross section of the volume common to the cylinders, enclosing a circular cross section of the sphere.

Think of all these plane sections as being packed together like the leaves of a book. Clearly, the volume of the sphere will be the sum of all circular cross sections, and the volume of the solid common to both cylinders will be the sum of all the square cross sections. We conclude, therefore that the ratio of the volume of the sphere to the volume of the solid common to the cylinders is the same as the ratio of the area of a circle to the area of a circumscribed square. A brief calculation shows that the latter ratio is $\pi/4$. This allows the following equation, in which x is the volume we seek:

$$\frac{4\pi r^3/3}{x} = \frac{\pi}{4}$$

The π's drop out, giving x a value of $16r^3/3$. The radius in this case is 1, so the volume common to both cylinders is 16/3. As Archimedes pointed out, it is exactly 2/3 the volume of a cube that encloses the sphere; that is, a cube with an edge equal to the diameter of each cylinder.

This solution is a famous application of what is known as "Cavalieri's principle" after Francesco Bonaventura Cavalieri (1598-1647), an Italian mathematical physicist and student of Galileo. In essence it says that solids, such as prisms, cones, cylinders, and pyramids which have the same height and corresponding cross sections of the same area, have the same volume. In proving his principle Cavalieri anticipated integral calculus by building up a volume by summing to a limit an infinite set of infinitesimal cross sections. The principle was known to Archimedes. In a lost book called *The Method* that was not found until 1906 (it is the book in which Archimedes answers the crossed cylinders problem), he attributes the principle to Democritus who used it for calculating the volume of a pyramid and a cone.

The problem can be generalized in various ways, notably to n mutually perpendicular cylinders of the same radius that intersect in spaces of three and higher dimensions. Even the case of three cylinders that intersect at right angles is beyond the power of Cavalieri's principle and requires the use of integral calculus. The volume common to three cylinders is $8r^3(2 - \sqrt{2})$.

The two-cylinder case has applications in architecture to what are called "barrel vaults" that form ceilings. When two such vaults intersect they create a cross vault—a portion of the surface of the solid shared by two crossed cylinders. There also are important applications in crystallography and engineering.

Another application of Cavalieri's principle is to an old brain teaser about a cork plug that will fit smoothly into three holes: a circle, square, and an almost equilateral triangle. (See Figure 27, left.) Put another way, what shape can, if turned properly, cast shadows that are circles, squares, and isosceles triangles?

The plug solving the problem is shown on the right. Assume its circular base has a radius of 1, its height 2, and the top edge, directly above a diameter of the base, is also 2. You can also think of the surface as generated by a straight line joining the sharp edge to the circumference of the base, and moving so that at all times it is parallel to a plane perpendicular to the sharp edge. Calculus will give the plug's volume, but Cavalieri's method again solves it more simply. All you need know is that the volume of a right circular cylinder is the area of its base times its altitude.

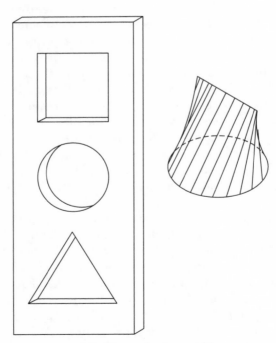

FIG. 27. The cork plug problem.

Here is how I gave the solution in my *Second Scientific American Book of Mathematical Puzzles and Diversions* (1961):

> Any vertical cross section of the cork plug at right angles to the top edge and perpendicular to the base will be a triangle. If the cork were a cylinder of the same height, corresponding cross sections would be rectangles. Each triangular cross section is obviously 1/2 the area of the corresponding rectangular cross section. Since all the triangular sections combine to make up the cylinder, the plug must be 1/2 the volume of the cylinder. The cylinder's volume is 2π, so our answer is simply π.
>
> Actually, the cork can have an infinite number of shapes and still fit the three holes. The shape described has the least volume of any convex solid that will fit the holes. The largest volume is obtained by the simple procedure of slicing the cylinder with two plane cuts as shown in Figure 28. This is the shape given in most puzzle books that include the plug problem. Its volume is equal to $2\pi - 8/3$.

Cavalieri's principle also applies to plane figures. If shapes between parallel lines *A* and *B*, as in Figure 29, have cross sections of the same area when cut by every line parallel to *A* and *B*, then they have the same area. As in the solid case, this generalizes to the theorem that if corresponding cross sections are always in the same ratio, their areas will have the same ratio.

CYCLOIDS AND HYPOCYCLOIDS

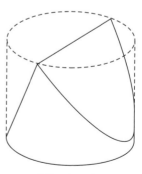

The curve generated by a point on a circle as it rolls along a straight line (see Figure 30) is called a cycloid. The length of the straight line segment from A to B is, of course, pi. Because pi is irrational, many mathematicians in days before calculus suspected that the length of the arc from cusp to cusp was also irrational. Today, almost all calculus textbooks show

FIG. 28. Another way to slice the cork plug.

Appendix

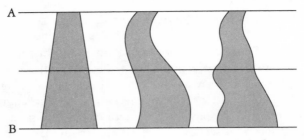

FIG. 29. Cavalieri's principle applied to plane figures.

how easy it is to determine that the arc's length is exactly four times the circle's diameter. It is also easy to determine by integration that the area below the arch is exactly three times the area of the circle. (For some of the cycloid's remarkable properties see Chapter 13 of my *Sixth Book of Mathematical Games,* 1971.)

Hypocycloids are closed curves generated by a point on a circle as it rolls inside a larger circle. They are discussed in many books on recreational mathematics as well as in calculus textbooks. Figure 31 shows two hypocycloids that have names. The deltoid is a three-cusped hypocycloid traced by a point on a rolling circle with a radius one-third or two-thirds that of the larger circle. The astroid of four cusps is produced by a point on a rolling circle one-fourth or three-fourths that of the containing circle.

A circle is not a graph of a function because a vertical line can cross it at two points, thereby giving two values to *y* for a given value of *x*. However, sectors of it can be integrated to obtain areas by using what are called "parametric" equations because they involve a third variable called their parameter. In the circle's case the parameter is the angle theta shown in Thompson's Figure 43, commonly designated by *t* or the Greek *θ*.

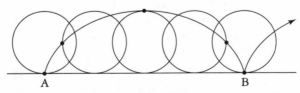

A   B

FIG. 30. How a cycloid is generated by a point on a rolling circle.

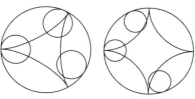

The Deltoid   The Astroid

FIG. 31.

If r is the radius of a circle centered at the graph's origin, its parametric equations are the trigonometric functions $x = r \cos \theta$, and $y = r \sin \theta$. Each value of theta, from 0 to 360°, determines a point on the circle.

Like the circle, the deltoid and astroid are not graphs of functions. Their areas are best obtained from their parametric equations based on the angle theta. Thompson does not discuss parametric equations, but you'll find them in modern textbooks. Integrating such equations gives the deltoid's area 2/9 the area of the larger circle, or $2\pi a^2$ where a is a rolling circle's radius that is 1/3 that of the larger circle. If that radius is 1, the deltoid's area is 2π. The astroid's area is 3/8 of its outer circle, and six times the rolling circle's area. The rolling circle has a radius 1/4 of the larger circle, or 2/3 times the rolling circle's area if the rolling circle has 3/4 the radius of the larger one.

I mention all this because of two connections with recreational geometry. What is the area of the two-cusped hypocycloid traced by a point on a rolling circle with half the radius of the larger circle? (See Figure 32) The surprising answer is zero! The hypocycloid is a straight line that is the large circle's diameter!

The minimum-size convex figure in which a needle (a straight line segment) of length 1 can be rotated 180 degrees is the equilateral triangle with a unit altitude and area of $\sqrt{3}/3$. What is the minimum *nonconvex* figure in which a unit needle can be rotated?

For many decades it was believed to be the deltoid. To everyone's surprise it was proved that the area could be as small as desired! I give the history of what is called "Kakeya's needle problem"

"Two-cusped"
Hypocycloid

FIG. 32.

in Chapter 18 of my *Unexpected Hanging and Other Mathematical Diversions* (1969). Hypocycloids and their sisters the epicycloids (generated by points on a circle that rolls *outside* another circle) are discussed in the first chapter of my *Wheels, Life, and Other Mathematical Amusements* (1983).

I have, of course, covered only a tiny fraction of calculus related problems that can be found in the literature of recreational mathematics. As you have seen, many calculus problems can be solved more easily by not using calculus. Indeed, there is a tendency of some mathematicians to look down their noses on non-calculus solutions to calculus problems, as though they were undignified tricks. On the contrary, they are just as useful and just as elegant as calculus solutions. The two classic references on this are: "No Calculus, Please," an entertaining paper by J. H. Butchart and Leo Moser in *Scripta Mathematica* (Vol. 18, September-December 1952, pp. 221-226), and Ivan Niven's *Maxima and Minima Without Calculus* (1981).

"The position taken by this book," Niven writes (p. 242), "is that, while calculus offers a powerful technique for solving extremal problems, there are other methods of great power that should not be overlooked. . . . . Many students . . . try to solve extremal questions . . . by seeking some function to differentiate, even though most of these problems are handled best by other methods. Moreover, students will often pursue the differentiation process through thick and thin, in spite of hopelessly complicated functions at hand."

Let me close this haphazard selection of problems by introducing a joke that physicist Richard Feynman once inflicted on fellow students at M.I.T. I quote from his autobiography *Surely You're Joking, Mr. Feynman* (1985):

> I often liked to play tricks on people when I was at MIT. One time, in mechanical drawing class, some joker picked up a French curve (a piece of plastic for drawing smooth curves—a curly, funny-looking thing) and said, "I wonder if the curves on this thing have some special formula?"
>
> I thought for a moment and said, "Sure they do. The curves are very special curves. Lemme show ya," and I picked up my French curve and began to turn it slowly.

"The French curve is made so that at the lowest point on each curve, no matter how you turn it, the tangent is horizontal."

All the guys in the class were holding their French curve up at different angles, holding their pencil up to it at the lowest point and laying it along, and discovering that, sure enough, the tangent is horizontal. They were all excited by this "discovery"—even though they had already gone through a certain amount of calculus and had already "learned" that the derivative (tangent) of the minimum (lowest point) of *any* curve is zero (horizontal). They didn't put two and two together. They didn't even know what they "knew."

INDEX

About the Authors

Silvanus Phillips Thompson was born in 1851, the son of a school teacher in York, England. He was President of the Institution of Electrical Engineers, was elected to the Royal Society in 1891, served as president of several scientific societies, and was the holder of many foreign honors and degrees.

Thompson wrote numerous technical books and manuals on electricity, magnetism, dynamos, and optics. He also wrote popular biographies of scientists Michael Faraday, Philipp Reis, and Lord Kelvin. As a devout Quaker and an active Knights Templar, he wrote two books about his faith: *The Quest for Truth* (1915) and *A Not Impossible Religion* (published posthumously in 1918).

Silvanus P. Thompson was also a popular lecturer and a skillful painter of landscapes. He died in 1916.

Born in 1914 in Tulsa, Oklahoma, the son of a petroleum geologist, Martin Gardner obtained a bachelor's degree in philosophy at the University of Chicago in 1936. Early jobs included reporter on the *Tulsa Tribune,* writer in the University of Chicago's office of press relations, and case worker in Chicago's Black Belt for the city's Relief Administration. After four years in the navy he began freelancing with sales of fiction to *Esquire.* In New York City he worked eight years as a contributing editor of *Humpty Dumpty's Magazine* before he began a twenty-five year stint with *Scientific American* as writer of that magazine's "Mathematical Games" column.

Gardner is now the author of some sixty books, most of them about mathematics and science. Among his awards are two honorary doctorates, and several prizes for math and science writing. He and his wife, Charlotte, live quietly in the mountains of North Carolina.